The Foundations of Scientific Inference

Toronto Nov 23, 1978

The Foundations of Scientific Inference

Wesley C. Salmon

University of Pittsburgh Press

ISBN 0-8229-5118-5
Library of Congress Catalog Card Number 67-21649

Manufactured in the United States of America
Second Printing, 1969
Third Printing, 1971
Fourth Printing, 1975

In memory of my teacher
Hans Reichenbach
(1891–1953)

Contents

The Foundations of Scientific Inference

> Let the course of things be allowed hitherto ever so regular; that alone, without some new argument or inference, proves not that, for the future, it will continue so. In vain do you pretend to have learned the nature of bodies from your past experience. Their secret nature, and consequently all their effects and influence, may change, without any change in their sensible qualities. This happens sometimes, and with regard to some objects: Why may it not happen always, and with regard to all objects? What logic, what process of argument secures you against this supposition? My practice, you say, refutes my doubts. But you mistake the purport of my question. As an agent, I am quite satisfied in the point; but as a philosopher, who has some share of curiosity, I will not say scepticism, I want to learn the foundation of this inference.
>
> —David Hume
>
> *An Enquiry Concerning Human Understanding*, § IV

Introduction

ALTHOUGH PERHAPS BORN EARLIER, mathematical physics came of age in the seventeenth century through the work of such men as Descartes, Galileo, Kepler, and Newton. This development constituted one of the most far-reaching of all revolutions in human thought, and it did not go unnoticed by various philosophers, some of whom had made significant contributions to it. There were, consequently, serious philosophic efforts to understand the logic of the new science.

Mathematical physics has an abstract and formal side as well as an observational and experimental side, and it has never been easy to understand the relations between them. Philosophies arising early in the modern period tended to emphasize one aspect to the neglect of the other. Descartes and Leibniz, impressed by the power of the mathematics they had helped to create, developed rationalistic epistemologies which failed to account for the observational and experimental aspect. Bacon and Locke, in their determination to handle the observational and experimental side, developed empirical theories of knowledge which failed to do justice to the abstract and formal character of science.

Although Descartes unequivocally rejected medieval authoritarianism with its unquestioning appeal to Aristotle, his conception of the logic of scientific knowledge was similar in fundamentals to that of Aristotle. Aristotle regarded scientific reasoning as strictly syllogistic in character; the only nonsyllogistic part is the establishment of first principles, and this is accomplished by intuitive induction. Intuitive induction is very different from inductive generalization as we think of it nowadays; it is, instead, a kind of rational insight.[1] For Descartes also, scientific knowledge consists of deduction from first principles established by the natural light of reason.[2]

Some of the salient differences between the rationalistic and empirical attitudes are expressed by Descartes' comments on Galileo in a letter to Father Mersenne, written in 1638:

> I find in general that he [Galileo] philosophizes much better than the average, in that he abandons as completely as he can the errors of the Schools, and attempts to examine physical matters by the methods of mathematics. In this I am in entire agreement with him, and I believe that there is absolutely no other way of discovering the truth. But it seems to me that he suffers greatly from continual digressions, and that he does not stop to explain all that is relevant to each point; which shows that he has not examined them in order, and that he has merely sought reasons for particular effects, without having considered the first causes of nature; and thus that he has built without a foundation. Indeed, because his fashion of philosophizing is so near to the truth, one can the more readily recognize his faults.[3]

While he applauds Galileo's use of mathematics, Descartes shows a complete lack of appreciation of the empirical approach. For Descartes, science is not an experimental enterprise in which one attempts to investigate clearly defined phenomena by observation of highly controlled situations. For him, the order is reversed. One understands the broadest aspects of nature by deduction from indubitable first principles; the details come at the end rather than the beginning. The first principles are grounded in pure reason. For the empiricist, on the other hand, the entire warrant for scientific theory rests upon its ability to explain precisely such details as can be handled experimentally.

It was Francis Bacon who first saw clearly that modern scientific method embodies a logic fundamentally different from that of Aristotle. In sharp contrast to Descartes, Bacon rejects rationalism and extols the method of careful observation and experimentation.

> There are and can be only two ways of searching into and discovering truth. The one flies from the senses and particulars to the most general axioms, and from these principles, the truth of which it takes for settled and immovable, proceeds to judgment and to the discovery of middle axioms. And this way is now in fashion. The other derives axioms from the senses and particulars,

rising by a gradual and unbroken ascent, so that it arrives at the most general axioms last of all. This is the true way, but as yet untried.[4]

Bacon realized that scientific knowledge must somehow be built upon inductive generalization from experience, and he tried to formulate the principles of this new logic—"a true induction." He confidently predicted that the assiduous application of this method would answer all important scientific questions. Looking back, we must regard his characterization as extremely primitive and wholly inadequate to the complexity of scientific method. His optimism for the future of science was charmingly naïve. He was, nevertheless, the enthusiastic herald of the new inductive method of science, and this in itself is an important contribution.

The seventeenth century could hardly imagine the foundational problems that were in store in connection with the methods of modern science. The spectacular successes of the newly developed methods led to an expansive attitude. The frontiers were pushed, but the foundations were seldom examined. Each of the previously noted aspects of mathematical physics had its foundational problems. The infinitesimal calculus was invented, and as an integral part of the mathematics of modern science it proved to be a powerful deductive tool. For about two centuries the foundations of calculus were in complete disorder.[5] The infinitesimal was a logical mystery; indeed, it was a logical absurdity. As a quantity smaller than any given positive quantity but greater than zero, it was a zero that was not really a zero—the ghost of a recently departed quantity! There was no clear understanding of mathematical functions, their values, or their limits. Illumination finally came early in the nineteenth century when Cauchy provided satisfactory explications of functions, limits, and derivatives. In the latter part of the century, further solidification resulted from the construction of the real number system and the arithmetization of calculus by Dedekind and Weierstrass. Moreover, even the logic of mathematical demonstration remained unexplicated until mathematical logic was invented and developed. Still, all was not well with the foundations. Logical analysis of mathematics pushed the notion of a class or set into greater and greater prominence, but even this apparently clear concept proved to be self-contradictory when defined in a natural way. At about the turn of the century, Russell derived his paradox of the class of all classes that do not belong to themselves from Cantor's theory of sets.[6]

Controversies and unsolved problems regarding the foundations of deductive inference still remain; foundations of mathematics is an active area of current research. Despite any remaining difficulties, however, there can be no question that our understanding of mathematics and deductive inference has been enormously enhanced by the investigations of the foundations. The foundations of inductive inference are far less fully understood, although they, too, are the object of considerable contemporary study. It is to the foundations of inductive inference in science that the present essay is devoted. I shall attempt to give a fairly well-rounded picture of the present state of investigations in this area, and in so doing I hope to present a convincing case for the desirability of continued study of these foundational questions.

The seventeenth century had even less basis for anticipating the foundational problems concerning inductive inference than it had for those of deductive inference. Although rationalists found various reasons for rejecting the empirical method, an awareness of the problem of induction cannot be counted as one of them. Ever since antiquity philosophers had been aware that the senses can deceive us, and this point was emphatically reiterated by Descartes. Those who were engaged in the quest for certainty found in this fact a basis for rejecting the empirical method. Some of the ancient skeptics even had been aware that inductive inference can sometimes lead to false conclusions; again, those engaged in the quest for certainty could reject inductive methods on this ground.[7] Philosophers who recognize that science cannot be expected to yield absolutely certain results can tolerate both of these shortcomings with equanimity. Hume's far more crushing critique of inductive inference came as a completely unanticipated development. It is notable that the most devastating criticism of empirical philosophy should come from a thoroughgoing empiricist. Hume's problem has proved astonishingly recalcitrant. Although there have been numerous efforts to solve or dissolve the difficulty, none is a satisfactory answer—at least, none is widely accepted as such.

Before turning to Hume, one other important seventeenth-century occurrence requires mention.[8] In 1654, the Chevalier de Méré, a gambler, asked Pascal to solve some problems relating to games of chance. One of these problems dealt with the fair division of stakes in a game not played through to completion; it is essentially the problem of calculating the probability that a player will win given a particular situation part way through the game. Another problem concerned the number of throws of two dice required to have at least a fifty-fifty chance of tossing a double six. Pascal communicated the problems to Fermat; in solving them, these two mathematicians founded the mathematical calculus of probability. Although originally applied only to gambling games, it was found to be applicable to wider and wider ranges of phenomena until today the concept of probability enjoys a fundamental position in all science. The development of probability theory provides still another example of foundational problems remaining long after the superstructure was well developed. For about two centuries it appears that the probability concept was regarded as easy to understand. To be sure, many authors presented definitions of the term, but this seems to have been a perfunctory gesture. The definitions were usually brief, and they were often quite out of harmony with the treatment of the subject when self-conscious definition was not under consideration. Nowhere can one find an appreciation of the severe difficulties encountered when detailed explication is undertaken. It is only with authors like Venn,[9] Peirce,[10] and Keynes[11] in the late nineteenth or early twentieth century that the problem of explicating the probability concept was taken seriously and treated at length. Furthermore, the relation between probability and induction was slow to be seen, and it is still widely misunderstood.

This, then, is the complex of problems I shall discuss: Hume's problem of induction, the problem of explicating the probability concept, and the problem of clarifying the relation between probability and inductive inference. Hume's critique of induction will constitute the historical as well as the logical point of departure.

I. The Problem of Induction

We all believe that we have knowledge of facts extending far beyond those we directly perceive. The scope of our senses is severely limited in space and time; our immediate perceptual knowledge does not reach to events that happened before we were born to events that are happening now in certain other places or to any future events. We believe, nevertheless, that we have some kind of indirect knowledge of such facts. We know that a glacier once covered a large part of North America, that the sun continues to exist at night, and that the tides will rise and fall tomorrow. Science and common sense have at least this one thing in common: Each embodies knowledge of matters of fact that are not open to our direct inspection. Indeed, science purports to establish general laws or theories that apply to all parts of space and time without restriction. A "science" that consisted of no more than a mere summary of the results of direct observation would not deserve the name.

Hume's profound critique of induction begins with a simple and apparently innocent question: How do we acquire knowledge of the unobserved?[12] This question, as posed, may seem to call for an empirical answer. We observe that human beings utilize what may be roughly characterized as inductive or scientific methods of extending knowledge from the observed to the unobserved. The sciences, in fact, embody the most powerful and highly developed methods known, and we may make an empirical investigation of scientific methods much as we might for any other sort of human behavior. We may consider the historical development of science. We may study the psychological, sociological, and political factors relevant to the pursuit of science. We may try to give an exact characterization of the behavior of scientists. In doing all these things, however, important and interesting as they are, we will have ignored the *philosophical* aspect of the problem Hume raised. Putting the matter very simply, these empirical investigations may enable us to describe the ways in which people arrive at *beliefs* about unobserved facts, but they leave open the question of whether beliefs arrived at in this way actually constitute *knowledge*. It is one thing to describe how people go about seeking to extend their knowledge; it is quite another to claim that the methods employed actually do yield knowledge.

One of the basic differences between knowledge and belief is that knowledge must be founded upon evidence—i.e., it must be belief founded upon some rational justification. To say that certain methods yield knowledge of the unobserved is to make a cognitive claim for them. Hume called into question the justification of such cognitive claims. The answer cannot be found entirely within an empirical study of human behavior, for a *logical* problem has been raised. It is the problem of understanding the logical relationship between evidence and conclusion in logically correct inferences. It is the problem of determining whether the inferences by which we attempt to make the transition from knowledge of the observed to knowledge of the unobserved are logically correct. The fact that people do or do not use a certain type of inference is irrelevant to its justifiability. Whether people have confidence in the correctness of a certain type of inference has nothing to do with whether such confidence is justified. If we should adopt a logically incorrect method for inferring one fact from others, these facts would not actually constitute evidence for the conclusion we have drawn. The problem of induction is the problem of explicating the very concept of *inductive evidence*.

There is another possibly misleading feature of the question as I have formulated it. When we ask how we can *acquire* knowledge of the unobserved, it sounds very much as if we are asking for a method for the *discovery* of new knowledge. This is, of course, a vital problem, but it is not the fundamental problem Hume raised. Whether there is or can be any sort of inductive logic of discovery is a controversial question I shall discuss in detail in a later section.[13] Leaving this question aside for now, there remains the problem of *justification* of conclusions concerning unobserved matters of fact. Given some conclusion, however arrived at, regarding unobserved facts, and given some alleged evidence to support that conclusion, the question remains whether that conclusion is, indeed, supported by the evidence offered in support of it.

Consider a simple and highly artificial situation. Suppose a number of balls have been drawn from an urn, and that all of the black ones that have been drawn are licorice-flavored. I am not now concerned with such psychological questions as what makes the observer note the color of these balls, what leads him to taste the black ones, what makes him take note of the fact that licorice flavor is associated with black color in his sample, or what makes him suppose that the black balls not yet drawn will also be licorice-flavored. The problem—Hume's basic *philosophical* problem—is this: Given that all of the observed black balls have been licorice-flavored, and given that somehow the conclusion has been entertained that the unobserved black balls in the urn are also licorice-flavored, do the observed facts constitute sound *evidence* for that conclusion? Would we be *justified* in accepting that conclusion on the basis of the facts alleged to be evidence for it?

As a first answer to this question we may point out that the inference does conform to an accepted inductive principle, a principle saying roughly that observed instances conforming to a generalization constitute evidence for it. It is, however, a very small step to the next question: What grounds have we for accepting this or any other inductive principle? Is there any reason or justification for placing confidence in the conclusions of inferences of this type? Given that the premises of this inference are true, and given that the inference conforms to a certain rule, can we provide any rational justification for accepting its conclusion rather than, for instance, the conclusion that black balls yet to be drawn will taste like quinine?

It is well known that Hume's answer to this problem was essentially skeptical. It was his great merit to have shown that a justification of induction, if possible at all, is by no means easy to provide. In order to appreciate the force of his argument it is first necessary to clarify some terminological points. This is particularly important because the word *induction* has been used in a wide variety of ways.

For purposes of systematic discussion one distinction is fundamental, namely, the distinction between demonstrative and nondemonstrative inference. A *demonstrative* inference is one whose premises necessitate its conclusion; the conclusion cannot be false if the premises are true. All valid deductions are demonstrative inferences. A *nondemonstrative* inference is simply one that fails to be demonstrative. Its conclusion is not necessitated by its premises; the conclusion could be false even if the premises are true. A demonstrative inference is *necessarily truth-preserving;* a nondemonstrative inference is not.

The category of nondemonstrative inferences, as I have characterized it, contains, among other things perhaps, all kinds of fallacious inferences. If, however, there is any kind of inference whose premises, although not necessitating the conclusion, do lend it weight, support it, or make it probable, then such inferences possess a certain kind of logical rectitude. It is not deductive validity, but it is important anyway. Inferences possessing it are *correct inductive inferences.*

Since demonstrative inferences have been characterized in terms of their basic property of necessary truth preservation, it is natural to ask how they achieve this very desirable trait. For a large group of demonstrative inferences, including those discussed under "valid deduction" in most logic texts, the answer is rather easy. Inferences of this type purchase necessary truth preservation by sacrificing any extension of content. The conclusion of such an inference says no more than do the premises—often less.[14] The conclusion cannot be false if the premises are true *because* the conclusion says nothing that was not already stated in the premises. The conclusion is a mere reformulation of all or part of the content of the premises. In some cases the reformulation is unanticipated and therefore psychologically surprising, but the conclusion cannot augment the content of the premises. Such inferences are *nonampliative;* an ampliative inference, then, has a conclusion with content not present either explicitly or implicitly in the premises.

While it is easy to understand why nonampliative inferences are necessarily truth-preserving, the further question arises whether there are any necessarily truth-preserving inferences that are also ampliative. Is there any type of inference whose conclusion must, of necessity, be true if the premises are true, but whose conclusion says something not stated by the premises? Hume believed that the answer is negative and so do I, but it is not easy to produce an adequate defense of this answer. Let us see, however, what an affirmative answer would amount to.

Suppose there were an ampliative inference that is also necessarily truth-preserving. Consider the implication from its premises, P_1,

. . . , P_k, to its conclusion C. If the inference were an ordinary nonampliative deduction, this implication would be analytic and empty; but since the argument is supposed to be ampliative, the implication must be synthetic. At the same time, because the argument is supposed to be necessarily truth-preserving, this implication must be not only true but necessarily true. Thus, to maintain that there are inferences that are both ampliative and necessarily truth-preserving is tantamount to asserting that there are synthetic a priori truths.[15] This may be seen in another way. Any ampliative inference can be made into a nonampliative one by adding a premise. In particular, if we add to the foregoing ampliative inference the synthetic a priori premise, "If P_1 and P_2 and . . . and P_k, then C," the resulting inference will be an ordinary valid nonampliative deduction. Consider our example once more; this time let us set it out more formally:

1. Some black balls from this urn have been observed.
   All observed black balls from this urn are licorice-flavored.

   All black balls in this urn are licorice-flavored.

This argument is clearly ampliative, for the premise makes a statement about observed balls only, while the conclusion makes a statement about the unobserved as well as the observed balls. It appears to be nondemonstrative as well, for it seems perfectly possible for the conclusion to be false even if the premises are true. We see no reason why someone might not have dropped a black marble in the urn which, when it is drawn, will be found to be tasteless. We could, however, rule out this sort of possibility by adding another premise:

2. Some black balls from this urn have been observed.
   All observed black balls in this urn are licorice-flavored.
   Any two balls in this urn that have the same color also have the same flavor.

   All black balls in this urn are licorice-flavored.

The additional premise has transformed the former nondemonstrative inference into a demonstrative inference, but we must also admit that we have transformed it into a nonampliative inference. If, however, the third premise of 2 were a synthetic a priori truth, the original inference, although ampliative, would have been necessarily truth-preserving and, hence, demonstrative. If the premise that transformed inference 1 into inference 2 were necessarily true, then it would be impossible for the conclusion of inference 1 to be false if the premises were true, for that would contradict the third premise of inference 2.

Hardly anyone would be tempted to say that the statement, "Any two balls in this urn that have the same color also have the same flavor," expresses a synthetic a priori truth. Other propositions have, however, been taken to be synthetic a priori. Hume and many of his successors noticed that typical inductive inferences, such as our example concerning licorice-flavored black balls, would seem perfectly sound if we could have recourse to some sort of principle of uniformity of nature. If we could only prove that the course of nature is uniform, that the future will be like the past, or that uniformities that have existed thus far will continue to hold in the future, then we would seem to be justified in generalizing from past cases to future cases—from the observed to the unobserved. Indeed, Hume suggests that we presuppose in our inductive reasoning a principle from which the third premise of 2 would follow as a special case: "We always presume, when we see like sensible qualities, that they have like secret powers, and expect that effects, similar to those which we have experienced, will follow from them."[16] Again, "From causes which appear *similar* we expect similar effects. This is the sum of all our experimental conclusions."[17]

Hume's searching examination of the principle of uniformity of nature revealed no ground on which it could be taken as a synthetic a priori principle. For all we can know a priori, Hume argued, the course of nature might change, the future might be radically unlike the past, and regularities that have obtained in respect to observed events might prove completely inapplicable to unobserved cases. We have found by experience, of course, that nature has exhibited a high degree of uniformity and regularity so far, and we infer inductively that this will continue, but to use an inductively inferred generalization as a justification for induction, as Hume emphasized, would be flagrantly circular. He concluded, in fact, that there are no synthetic a priori principles in virtue of which we could have demonstrative inferences that are ampliative. Hume recognized two kinds of reasoning: reasoning concerning relations of ideas and reasoning concerning matters of fact and existence. The former is demonstrative but nonampliative while the latter is ampliative but not necessarily truth-preserving.

If we agree that there are no synthetic a priori truths, then we must identify necessarily truth-preserving inference with nonampliative inference. All ampliative inference is nondemonstrative. This leads to an exhaustive trichotomy of inferences: valid deductive inference, correct inductive inference, and assorted fallacies. The first question is, however, whether the second category is empty or whether there are such things as correct inductive inferences. This is Hume's problem of induction. Can we show that any particular type of ampliative inference can be justified in any way? If so, it will qualify as correct induction.

Consider, then, any ampliative inference whatever. The example of the licorice-flavored black balls illustrates the point. We cannot show *deductively* that this inference will have a true conclusion given true premises. If we could, we would have proved that the conclusion must be true if the premises are. That would make it necessarily truth-preserving, hence, demonstrative. This, in turn, would mean that it was nonampliative, contrary to our hypothesis. Thus, if an ampliative inference could be justified deductively it would not be ampliative. It follows that ampliative inference cannot be justified deductively.

At the same time, we cannot justify any sort of ampliative inference *inductively*. To do so would require the use of some sort of nondemonstrative inference. But the question at issue is the justification of nondemonstrative inference, so the procedure would be question begging. Before we can properly employ a nondemonstrative inference in a justifying argument, we must already have justified that nondemonstrative inference.

Hume's position can be summarized succinctly: We cannot justify any kind of ampliative inference. If it could be justified deductively it would not be ampliative. It cannot be justified nondemonstratively because that would be viciously circular. It seems, then, that there is no way in which we can extend our knowledge to the unobserved. We have, to be sure, many beliefs about the unobserved, and in some of them we place great confidence. Nevertheless, they are without rational justification of any kind!

This is a harsh conclusion, yet it seems to be supported by impeccable arguments. It might be called "Hume's paradox," for the conclusion, although ingeniously argued, is utterly repugnant to common sense and our deepest convictions. We *know* ("in our hearts") that we have knowledge of unobserved fact. The challenge is to show how this is possible.

II. Attempted Solutions

It hardly needs remarking that philosophers have attempted to meet Hume's intriguing challenge in a wide variety of ways. There have been direct attacks upon some of Hume's arguments. Attempts to provide inductive arguments to support induction and attempts to supply a synthetic a priori principle of uniformity of nature belong in this category. Some authors have claimed that the whole problem arises out of linguistic confusion, and that careful analysis shows it to be a pseudoproblem. Some have even denied that inductive inference is needed, either in science or in everyday affairs. In this section I shall survey what seem to me to be the most important efforts to deal with the problem.

1. Inductive Justification. If Hume's arguments had never been propounded and we were asked why we accept the methods of science, the most natural answer would be, I think, that these methods have proved themselves by their results. We can point to astonishing technological advances, to vastly increased comprehension, and to impressive predictions. Science has provided us with foresight, control, and understanding. No other method can claim a comparable record of successful accomplishment. If methods are to be judged by their fruits, there is no doubt that the scientific method will come out on top.

Unfortunately, Hume examined this argument and showed that it is viciously circular. It is an example of an attempt to justify inductive methods inductively. From the premise that science has had considerable predictive success in the past, we conclude that it will continue to have substantial predictive success in the future. Observed cases of the application of scientific method have yielded successful prediction; therefore, as yet unobserved cases of the application of scientific method will yield successful predictions. This argument has the same structure as our black-balls-in-the-urn example; it is precisely the sort of ampliative inference from the observed to the unobserved whose justifiability is in question.

Consider the parallel case for a radically different sort of method. A crystal gazer claims that his method is the appropriate method for making predictions. When we question his claim he says, "Wait a moment; I will find out whether the method of crystal gazing is the best method for making predictions." He looks into his crystal ball and announces that future cases of crystal gazing will yield predictive success. If we should protest that his method has not been especially successful in the past, he might well make certain remarks about parity of reasoning. "Since you have used your method to justify your method, why shouldn't I use my method to justify my method? If you insist upon judging my method by using your method, why shouldn't I use my method to evaluate your method? By the way, I note by gazing into my crystal ball that the scientific method is now in for a very bad run of luck."

The trouble with circular arguments is obvious: with an appropriate circular argument you can prove anything. In recent years, nevertheless, there have been several notable attempts to show how inductive rules can be supported inductively. The authors of such attempts try to show, of course, that their arguments are not circular. Although they argue persuasively, it seems to me that they do not succeed in escaping circularity.

One of the most widely discussed attempts to show that self-supporting inductive inferences are possible without circularity is due to Max Black.[18] Black correctly observes that the traditional fallacy of circular argument (*petitio principii*) involves assuming as a premise, often unwittingly, the conclusion that is to be proved. Thus, for example, a variety of "proofs" of Euclid's fifth postulate offered by mathematicians for about two millenia before the discovery of non-Euclidean geometry are circular in the standard fashion. They fail to show that the fifth postulate follows from the first four postulates alone; instead, they require in addition the assumption of a proposition equivalent to the proposition being demonstrated. The situation is quite different for self-supporting inductive arguments. The conclusion to be proved does not appear as one of the premises. Consider one of Black's examples: [19]

3. In most instances of the use of R_2 in arguments with true premises examined in a wide variety of conditions, R_2 has usually been successful.

   Hence (probably):

   In the next instance to be encountered of the use of R_2 in an argument with true premises, R_2 will be successful.

To say that an argument with true premises is successful is merely to say that it has a true conclusion. The rule R_2 is

> To argue from *Most instances of A's examined in a wide variety of conditions have been B* to (probably) *The next A to be encountered will be B.*

Inference 3 can be paraphrased suggestively, although somewhat inaccurately, as:

4. R_2 has usually been successful in the past.

   Hence (probably):

   R_2 will be successful in the next instance.

Inference 3 is governed by R_2, that is, it conforms to the stipulation laid down by R_2. R_2 is *not* a premise, however, nor is any statement to the effect that all, some, or any future instances of R_2 will be successful. As Lewis Carroll showed decisively, there is a fundamental distinction between premises and rules of inference.[20] Any inference, inductive or deductive, must conform to some rule, but neither the rule nor any statement about the rule is to be incorporated into the inference as an additional premise. If such additional premises were required, inference would be impossible. Thus, inference 3 is not a standard *petitio principii.*

What, then, are the requirements for a self-supporting argument? At least three are immediately apparent: (1) The argument must have true premises. (2) The argument must conform to a certain rule. (3) The conclusion of that argument must say something about the success or reliability of that rule in unexamined instances of its application. Inference 3 has these characteristics.

It is not difficult to find examples of deductive inferences with the foregoing characteristics.

5. If snow is white, then *modus ponens* is valid.
   Snow is white.

   Modus ponens is valid.

Inference 5 may seem innocuous enough, but the same cannot be said for the following inference:

6. If affirming the consequent is valid, then coal is black.
   Coal is black.

   Affirming the consequent is valid.

Like inference 5, inference 6 has true premises, it conforms to a certain rule, and its conclusion asserts the validity of that rule. Inference 5 did nothing to enhance our confidence in the validity of *modus ponens,* for we have far better grounds for believing it to be valid. Inference 6 does nothing to convince us that affirming the consequent is valid, for we know on other grounds that it is invalid. Arguments like 5 and 6 are, nevertheless, instructive. Both are circular in some sense, though neither assumes *as a premise* the conclusion it purports to establish. In deductive logic the situation is quite straightforward. A deductive inference establishes its conclusion if it has true premises and has a valid form. If either of these features is lacking the conclusion is not established by that argument. If the argument is valid but the premises are not true we need not accept the conclusion. If the premises are true but the argument is invalid we need not accept the conclusion. One way in which an argument can be circular is by adopting as a premise the very conclusion that is to be proved; this is the fallacy of *petitio principii* which I shall call "premise-circularity." Another way in which an argument can be circular is by exhibiting a form whose validity is asserted by the very conclusion that is to be proved; let us call this type of circularity "rule-circularity." Neither type of circular argument establishes its conclusion in any interesting fashion, for in each case the conclusiveness of the argument depends upon the assumption of the conclusion of that argument. Inferences 5 and 6 are not premise-circular; each is rule-circular. They are, nevertheless, completely question begging.

The situation in induction is somewhat more complicated, but basically the same.[21] Consider the following argument:

7. In most instances of the use of R_3 in arguments with true premises examined in a wide variety of conditions, R_3 has usually been *un*successful.

   Hence (probably):

   In the next instance to be encountered of the use of R_3 in an argument with true premises, R_3 will be successful.

The rule R_3 is

> To argue from *Most instances of A's examined in a wide variety of conditions have been non-B* to (probably) *The next A to be encountered will be B.*

Inference 7 can be paraphrased as follows:

8. R_3 has usually been unsuccessful in the past.

   Hence (probably):

   R_3 will be successful in the next instance.

Notice that there is a perfect parallel between R_2, 3, 4 on the one hand and R_3, 7, 8 on the other. Since those instances in which R_2 would be successful are those in which R_3 would be unsuccessful, the premises of 3 and 4 describe the same state of affairs as do the premises of 7 and 8. Thus, the use of R_3 in the next instance seems to be supported in the same manner and to the same extent as the use of R_2 in the next instance. However, R_2 and R_3 conflict directly with each other. On the evidence that most Italians examined in a wide variety of conditions have been dark-eyed, R_2 allows us to infer that the next Italian to be encountered will be dark-eyed, while R_3 permits us to infer from the same evidence that he will have light-colored eyes. It appears then that we can construct self-supporting arguments for correct and incorrect inductive rules just as we can for valid and invalid deductive rules.

Black would reject self-supporting arguments for the fallacy of affirming the consequent and for a counterinductive rule like R_3, because we know on independent grounds that such rules are faulty. Affirming the consequent is known to be fallacious, and the counterinductive method can be shown to be self-defeating. An additional requirement for a self-supporting argument is that the rule thus supported be one we have no independent reason to reject. Nevertheless, the fact that we can construct self-supporting arguments for such rules should give us pause. What if we had never realized that affirming the consequent is fallacious? What if we had never noticed anything wrong with the counterinductive method? Would arguments like 6, 7, and 8 have to be considered cogent? What about the standard inductive method? Is it as incorrect as the counterinductive method, but for reasons most of us have not yet realized?

It sounds as if a self-supporting argument is applicable only to rules we already know to be correct; as a matter of fact, this is the view Black holds. He has argued in various places that induction is in no need of a general justification.[22] He holds that calling into question of all inductive methods simultaneously results in a hopelessly skeptical position. He is careful to state explicitly at the outset of his discussion of self-supporting inductive arguments that he is not dealing with the view "that *no* inductive argument ought to be regarded as correct until a philosophical justification of induction has been provided."[23] At the conclusion he acknowledges, moreover, that "anybody who thinks he has good grounds for condemning all inductive arguments will also condemn inductive arguments in support of inductive rules."[24] Black is careful to state explicitly that self-supporting inductive arguments provide no answer to the problem of justification of induction as raised by Hume. What good, then, are self-supporting inductive arguments?

In deductive logic, correctness is an all-or-nothing affair. Deductive inferences are either totally valid or totally invalid; there cannot be such a thing as degree of validity. In inductive logic the situation is quite different. Inductive correctness does admit of degrees; one inductive conclusion may be more strongly supported than another. In this situation it is possible, Black claims, to have an inductive rule we know to be correct to some degree, but whose status can be enhanced by self-supporting arguments. We might think a rather standard inductive rule akin to Black's R_2 is pretty good, but through inductive investigation of its application we might find that it is extremely good—much better than we originally thought. Moreover, the inductive inferences we use to draw that conclusion might be governed by precisely the sort of rule we are investigating. It is also possible, of course, to find by inductive investigation that the rule is not as good as we believed beforehand.

It is actually irrelevant to the present discussion to attempt to evaluate Black's view concerning the possibility of increasing the justification of inductive rules by self-supporting arguments. The important point is to emphasize, because of the possibility of constructing self-supporting arguments for counterinductive rules, that the attempt to provide inductive support of inductive rules cannot, without vicious circularity, be applied to the problem of justifying induction from scratch. If there is any way of providing the beginnings of a justification, or if we could show that some inductive rule stands in no need of justification in the first instance, then it would be suitable to return to Black's argument concerning the increase of support. I am not convinced, however, that Black has successfully shown that there is a satisfactory starting place.

I have treated the problem of inductive justification of induction at some length, partly because other authors have not been as cautious as Black in circumscribing the limits of inductive justification of induction.[25] More important, perhaps, is the fact that it is extremely difficult, psychologically speaking, to shake the view that past success of the inductive method constitutes a genuine justification of induction. Nevertheless, the basic fact remains: Hume showed that inductive justifications of induction are fallacious, and no one has since proved him wrong.

2. *The Complexity of Scientific Inference.* The idea of a philosopher discussing inductive inference in science is apt to arouse grotesque images in many minds. People are likely to imagine someone earnestly attempting to explain why it is reasonable to conclude that the sun will rise tomorrow morning because it always has done so in the past. There may have been a time when primitive man anticipated the dawn with assurance based only upon the fact that he had seen dawn follow the blackness of night as long as he could remember, but this primitive state of knowledge, if it ever existed, was unquestionably *pre*scientific. This kind of reasoning bears no resemblance to science; in fact, the crude induction exhibits a complete absence of scientific understanding. Our scientific reasons for believing that the sun will rise tomorrow are of an entirely different kind. We understand the functioning of the solar system in terms of the laws of physics. We predict particular astronomical

occurrences by means of these laws in conjunction with a knowledge of particular initial conditions that prevail. Scientific laws and theories have the logical form of general statements, but they are seldom, if ever, simple generalizations from experience.

Consider Newton's gravitational theory: Any two bodies are mutually attracted by a force proportional to the product of their masses and inversely proportional to the square of the distance between their centers. Although general in form, this kind of statement is not established by generalization from instances. We do not go around saying, "Here are two bodies—the force between them is such and such; here are two more bodies—the force between them is such and such; etc." Scientific theories are taken quite literally as hypotheses. They are entertained in order that their consequences may be drawn and examined. Their acceptability is judged in terms of these consequences. The consequences are extremely diverse—the greater the variety the better. For Newtonian theory, we look to such consequences as the behavior of Mars, the tides, falling bodies, the pendulum, and the torsion balance. These consequences have no apparent unity among themselves; they do not constitute a basis for inductive generalization. They achieve a kind of unity only by virtue of the fact that they are consequences of a single physical theory.

The type of inference I have been characterizing is very familiar; it is known as the *hypothetico-deductive method.*[26] It stands in sharp contrast to *induction by enumeration,* which consists in simple inductive generalization from instances. Schematically, the hypothetico-deductive method works as follows: From a general hypothesis and particular statements of initial conditions, a particular predictive statement is deduced. The statements of initial conditions, at least for the time, are accepted as true; the hypothesis is the statement whose truth is at issue. By observation we determine whether the predictive statement turned out to be true. If the predictive consequence is false, the hypothesis is disconfirmed. If observation reveals that the predictive statement is true, we say that the hypothesis is confirmed to some extent. A hypothesis is not, of course, conclusively proved by any one or more positively confirming instances, but it may become highly confirmed. A hypothesis that is sufficiently confirmed is accepted, at least tentatively.

It seems undeniable that science uses a type of inference at least loosely akin to the hypothetico-deductive method.[27] This has led some people to conclude that the logic of science is thoroughly deductive in character. According to this view, the only nondeductive aspect of the situation consists in thinking up hypotheses, but this is not a matter of logic and therefore requires no justification. It is a matter of psychological ingenuity of discovery. Once the hypothesis has been discovered, by some entirely nonlogical process, it remains only to *deduce* consequences and check them against observation.

It is, of course, a fallacy to conclude that the premises of an argument must be true if its conclusion is true. This fact seems to be the basis for the quip that a logic text is a book that consists of two parts; in the first part (on deduction) the fallacies are explained, in the second part (on induction) they are committed. The whole trouble with saying that the hypothetico-deductive method renders the logic of science entirely deductive is that we are attempting to establish a *premise* of the deduction, not the conclusion. Deduction is an indispensible part of the logic of the hypothetico-deductive method, but it is not the only part. There is a fundamental and important sense in which the hypothesis must be regarded as a conclusion instead of a premise. Hypotheses (later perhaps called "theories" or "laws") are among the *results* of scientific investigation; science aims at establishing general statements about the world. Scientific prediction and explanation require such generalizations. While we are concerned with the status of the general hypothesis—whether we should accept it or reject it—the hypothesis must be treated as a conclusion to be supported by evidence, not as a premise lending support to other conclusions. The inference *from* observational evidence *to* hypothesis is surely not deductive. If this point is not already obvious it becomes clear the moment we recall that for any given body of observational data there is, in general, more than one hypothesis compatible with it. These alternative hypotheses differ in factual content and are incompatible with each other. Therefore, they cannot be deductive consequences of any consistent body of observational evidence.

We must grant, then, that science embodies a type of inference resembling the hypothetico-deductive method and fundamentally different from induction by enumeration. Hume, on the other hand, has sometimes been charged with a conception of science according to which the only kind of reasoning is induction by enumeration. His typical examples are cases of simple generalization of observed regularities, something like our example of the licorice-flavored black balls. In the past, water has quenched thirst; in the future, it will as well. In the past, fires have been hot; in the future, they will be hot. In the past, bread has nourished; in the future, it will do so likewise. It might be said that Hume, in failing to see the essential role of the hypothetico-deductive method, was unable to appreciate the complexity of the theoretical science of his own time, to say nothing of subsequent developments. This is typical, some might say, of the misunderstandings engendered by philosophers who undertake to discuss the logic of science without being thoroughly conversant with mathematics and natural science.

This charge against Hume (and other philosophers of induction) is ill-founded. It was part of Hume's genius to have recognized that the arguments he applied to simple enumerative induction apply equally to any kind of ampliative or nondemonstrative inference whatever. Consider the most complex kind of scientific reasoning—the most elaborate example of hypothetico-deductive inference you can imagine. Regardless of subtle features or complications, it is ampliative overall. The conclusion is a statement whose content exceeds the observational evidence upon which it is based. A scientific theory that merely summarized what had already been observed would not deserve to be called a theory. If scientific inference were not ampliative, science would be useless for prediction, postdiction, and explanation. The highly general results that are the pride of theoretical science would be impossible if scientific inference were not ampliative.

In presenting Hume's argument, I was careful to set it up so that it would apply to any kind of ampliative or nondemonstrative inference, no matter how simple or how complex. Furthermore, the distinction between valid deduction and nondemonstrative inference is completely exhaustive. Take any inference whatsoever. It must be deductive or nondemonstrative. Suppose it is nondemonstrative. If we could justify it deductively it would cease to be nondemonstrative. To justify it nondemonstratively would presuppose an already justified type of nondemonstrative inference, which is precisely the problem at issue. Hume's argument does *not* break down when we consider forms more complex than simple enumeration. Although the word "induction" is sometimes used as a synonym for "induction by simple enumeration," I am not using it in that way. Any type of logically correct ampliative inference is induction; the problem of induction is to show that some particular form of ampliative inference is justifiable. It is in this sense that we are concerned with the problem of the justification of inductive inference.

A further misunderstanding is often involved in this type of criticism of Hume. There is a strong inclination to suppose that induction is regarded as the method by which scientific results are discovered.[28] Hume and other philosophers of induction are charged with the view that science has developed historically through patient collection of facts and generalization from them. I know of no philosopher—not even Francis Bacon!—who has held this view, although it is frequently attacked in the contemporary literature.[29] The term "generalization" has an unfortunate ambiguity which fosters the confusion. In one meaning, "generalization" refers to an inferential process in which one makes a sort of mental transition from particulars to a universal proposition; in this sense, generalization is an act of generalizing—a process that yields general results. In another meaning, "generalization" simply refers to a universal type of proposition, without any reference to its source or how it was thought of. It is entirely possible for science to contain many generalizations (in the latter sense) without embodying any generalizations (in the former sense). As I said explicitly at the outset, the problem of induction I am discussing is a problem concerning justification, not discovery. The thesis I am defending—that science does embody induction in a logically indispensable fashion—has nothing to do with the history of science or the psychology of particular scientists. It is simply the claim that scientific inference is ampliative.

3. *Deductivism.* One of the most interesting and controversial contemporary attempts to provide an account of the logic of science is Karl Popper's deductivism.[30] In the preceding section I discussed the view that the presence of the hypothetico-deductive method in the logic of science makes it possible to dispense with induction in science and, thereby, to avoid the problem of induction. I argued that the hypothetico-deductive method, since it is ampliative and nondemonstrative, is not strictly deductive; it is, in fact, inductive in the relevant sense. As long as the hypothetico-deductive method is regarded as a method for supporting scientific hypotheses, it cannot succeed in making science thoroughly deductive. Popper realizes this, so in arguing that deduction is the sole mode of inference in science he rejects the hypothetico-deductive method as a means for confirming scientific hypotheses. He asserts that induction plays no role whatever in science; indeed, he maintains that there is no such thing as correct inductive inference. Inductive logic is, according to Popper, a complete delusion. He admits the psychological fact that people (including himself) have faith in the uniformity of nature, but he holds, with Hume, that this can be no more than a matter of psychological fact. He holds, with Hume, that there can be no rational justification of induction, and he thinks Hume proved this point conclusively.

Popper's fundamental thesis is that falsifiability is the mark by which statements of empirical science are distinguished from metaphysical statements and from tautologies. The choice of falsifiability over verifiability as the criterion of demarcation is motivated by a long familiar fact—namely, it is possible to falsify a universal generalization by means of one negative instance, while it is impossible to verify a universal generalization by any limited number of positive instances. This, incidentally, is the meaning of the old saw which is so often made into complete nonsense: "The exception proves the rule." In this context, a rule is a universal generalization, and the term "to prove" means archaically "to test." The exception (i.e., the negative instance) proves (i.e., tests) the rule (i.e., the universal generalization), not by showing it to be true, but by showing it to be false. There is no kind of positive instance to prove (i.e., test) the rule, for positive instances are completely indecisive. Scientific hypotheses, as already noted, are general in form, so they are amenable to falsification but not verification.

Popper thus holds that falsifiability is the hallmark of empirical science. The aim of empirical science is to set forth theories to stand the test of every possible serious attempt at falsification. Scientific theories are hypotheses or conjectures; they are general statements designed to explain the world and make it intelligible, but they are never to be regarded as final truths. Their status is always that of tentative conjecture, and they must continually face the severest possible criticism. The function of the theoretician is to propose scientific conjectures; the function of the experimentalist is to devise every possible way of falsifying these theoretical hypotheses. The attempt to confirm hypotheses is no part of the aim of science.[31]

General hypotheses by themselves do not entail any predictions of particular events, but they do in conjunction with statements of initial conditions. The laws of planetary motion in conjunction with statements about the relative positions and velocities of the earth, sun, moon, and planets enable us to predict a solar eclipse. The mode of inference is deduction. We have a high degree of intersubjective agreement concerning the initial conditions, and we likewise can obtain intersubjective agreement as to whether the sun's disc was obscured at the predicted time and place. If the predicted fact fails to occur, the theory has suffered falsification. Again, the mode of inference is deduction. If the theory were true, then, given the truth of the statements of initial conditions, the prediction would have to be true. The prediction, as it happens, is false; therefore, the theory is false. This is the familiar principle of *modus tollens;* it is, according to Popper, the only kind of inference available for the acceptance or rejection of hypotheses, and it is clearly suitable for rejection only.

Hypothetico-deductive theorists maintain that we have a confirming instance for the theory if the eclipse occurs as predicted. Confirming instances, they claim, tend to enhance the probability of the hypothesis or give it inductive support. With enough confirming instances of appropriate kinds, the probability of the hypothesis becomes great enough to warrant accepting it as true—not, of course, with finality and certainty, but provisionally. With sufficient inductive support of this kind we are justified in regarding it as well established. Popper, however, rejects the positive account, involving as it does the notion of inductive support. If a hypothesis is tested and the result is negative, we can reject it. If the test is positive, all we can say is that we have failed to falsify it. We cannot say that it has been confirmed or that it is, because of the positive test result, more probable. Popper does admit a notion of *corroboration* of hypotheses, but that is quite distinct from confirmation. We shall come to corroboration presently. For the moment, all we have are successful or unsuccessful attempts at falsification; all we can say about our hypotheses is that they are falsified or unfalsified. This is as far as inference takes us; according to Popper, this is the limit of logic. Popper therefore rejects the hypothetico-deductive method as it is usually characterized and accepts only the completely deductive *modus tollens.*

Popper—quite correctly I believe—denies that there are absolutely basic and incorrigible protocol statements that provide the empirical foundation for all of science. He does believe that there are relatively basic observation statements about macroscopic physical occurrences concerning which we have a high degree of intersubjective agreement. Normally, we can accept as unproblematic such statements as, "There is a wooden table in this room," "The pointer on this meter stands between 325 and 350," and "The rope just broke and the weight fell to the floor." Relatively basic statements of this kind provide the observation base for empirical science. This is the stuff of which empirical tests of scientific theories are made.

Although Popper's basic statements must in the last analysis be considered hypotheses, falsifiable and subject to test like other scientific hypotheses, it is obvious that the kinds of hypotheses that constitute theoretical science are far more general than the basic statements. But now we must face the grim fact that valid deductive inference, although necessarily truth-preserving, is nonampliative.[32] It is impossible to deduce from accepted basic statements any conclusion whose content exceeds that of the basic statements themselves. Observation statements and deductive inference yield nothing that was not stated by the observation statements themselves. If science consists solely of observation statements and deductive inferences, then talk about theories, their falsifiability, and their tests is empty. The content of science is coextensive with the content of the statements used to describe what we directly observe. There are no general theories, there is no predictive content, there are no inferences to the remote past. Science is barren.

Consider a few simple time-honored examples. Suppose that the statement "All ravens are black" has been entertained critically and subjected to every attempt at falsification we can think of. Suppose it has survived all attempts at falsification. What is the scientific content of all this? We can say that "All ravens are black" has not been falsified, which is equivalent to saying that we have not observed a nonblack raven. This statement is even poorer in content than a simple recital of our color observations of ravens. To say that the hypothesis has not been falsified is to say less than is given in a list of our relevant observation statements. Or, consider the generalization, "All swans are white." What have we said when we say that this hypothesis has been falsified? We have said only that a nonwhite swan has been found. Again, the information conveyed by this remark is less than we would get from a simple account of our observations of swans.

Popper has never claimed that falsification by itself can establish scientific hypotheses. When one particular hypothesis has been falsified, many alternative hypotheses remain unfalsified. Likewise, there is nothing unique about a hypothesis that survives without being falsified. Many other unfalsified hypotheses remain to explain the same facts. Popper readily admits all of this. If science is to amount to more than a mere collection of our observations and various reformulations thereof, it must embody some other methods besides observation and deduction. Popper supplies that additional factor: *corroboration.*[33]

When a hypothesis has been falsified, it is discarded and replaced by another hypothesis which has not yet experienced falsification. Not all unfalsified hypotheses are on a par. There are principles of selection among unfalsified hypotheses. Again, falsifiability is the key. Hypotheses differ from one another with respect to the ease with which they can be falsified, and we can often compare them with respect to degree of falsifiability. Popper directs us to seek hypotheses that are as highly falsifiable as possible. Science, he says, is interested in bold conjectures.

These conjectures must be consistent with the known facts, but they must run as great a risk as possible of being controverted by the facts still to be accumulated. Furthermore, the search for additional facts should be guided by the effort to find facts that will falsify the hypothesis.

As Popper characterizes falsifiability, the greater the degree of falsifiability of a hypothesis, the greater its content. Tautologies lack empirical content because they do not exclude any possible state of affairs; they are compatible with any possible world. Empirical statements are not compatible with every possible state of affairs; they are compatible with some and incompatible with others. The greater the number of possible states of affairs excluded by a statement, the greater its content, for the more it does to pin down our actual world by ruling out possible but nonactual states of affairs. At the same time, the greater the range of facts excluded by a statement—the greater the number of situations with which the statement is incompatible—the greater the risk it runs of being false. A statement with high content has more *potential falsifiers* than a statement with low content. For this reason, high content means high falsifiability. At the same time, content varies inversely with probability. The logical probability of a hypothesis is defined in terms of its range—that is, the possible states of affairs with which it is compatible. The greater the logical probability of a hypothesis, the fewer are its potential falsifiers. Thus, high probability means low falsifiability.

Hypothetico-deductive theorists usually recommend selecting, from among those hypotheses that are compatible with the available facts, the most probable hypothesis. Popper recommends the opposite; he suggests selecting the most falsifiable hypothesis. Thus, he recommends selecting a hypothesis with low probability. According to Popper, a highly falsifiable hypothesis which is severely tested becomes highly corroborated. The greater the severity of the tests—the greater their number and variety—the greater the corroboration of the hypothesis that survives them.

Popper makes it very clear that hypotheses are not regarded as true because they are highly corroborated. Hypotheses cannot be firmly and finally established in this or any other way. Furthermore, because of the inverse relation between falsifiability and probability, we cannot regard highly corroborated hypotheses as probable. To be sure, a serious attempt to falsify a hypothesis which fails does add to the corroboration of this hypothesis, so there is some similarity between corroboration and confirmation as hypothetico-deductive theorists think of it, but it would be a misinterpretation to suppose that increasing corroboration is a process of accumulating positive instances to increase the probability of the hypothesis.[34]

Nevertheless, Popper does acknowledge the need for a method of selecting among unfalsified hypotheses. He has been unequivocal in his emphasis upon the indispensability of far-reaching theory in science. Empirical science is not an activity of merely accumulating experiences; it is theoretical through and through. Although we do not regard any hypotheses as certainly true, we do accept them tentatively and provisionally. Highly corroborated hypotheses are required for prediction and explanation. From among the ever-present multiplicity of hypotheses compatible with the available evidence, we select and accept.

There is just one point I wish to make here regarding Popper's theory. It is not properly characterized as *deductivism.* Popper has not succeeded in purging the logic of science of all inductive elements. My reason for saying this is very simple. Popper furnishes a method for selecting hypotheses whose content exceeds that of the relevant available basic statements. Demonstrative inference cannot accomplish this task alone, for valid deductions are nonampliative and their conclusions cannot exceed their premises in content. Furthermore, Popper's theory does not pretend that basic statements plus deduction can give us scientific theory; instead, corroboration is introduced. Corroboration is a nondemonstrative form of inference. It is a way of providing for the acceptance of hypotheses even though the content of these hypotheses goes beyond the content of the basic statements. *Modus tollens* without corroboration is empty; *modus tollens* with corroboration is induction.

When we ask, "Why should we reject a hypothesis when we have accepted one of its potential falsifiers?" the answer is easy. The potential falsifier contradicts the hypothesis, so the hypothesis is false if the potential falsifier holds. That is simple deduction. When we ask, "Why should we accept from among all the unfalsified hypotheses one that is highly corroborated?" we have a right to expect an answer. The answer is some kind of justification for the methodological rule—for the method of corroboration. Popper attempts to answer this question.

Popper makes it clear that his conception of scientific method differs in important respects from the conceptions of many inductivists. I do not want to quibble over a word in claiming that Popper is, himself, a kind of inductivist. The point is not a trivial verbal one. Popper has claimed that scientific inference is exclusively deductive. We have seen, however, that demonstrative inference is not sufficient to the task of providing a reconstruction of the logic of the acceptance—albeit tentative and provisional—of hypotheses. Popper himself realizes this and introduces a mode of nondemonstrative inference. It does not matter whether we call this kind of inference "induction"; whatever we call it, it is ampliative and not necessarily truth preserving. Using the same force and logic with which Hume raised problems about the justification of induction, we may raise problems about the justification of any kind of nondemonstrative inference. As I argued in the preceding section, Hume's arguments are not peculiar to induction by enumeration or any other special kind of inductive inference; they apply with equal force to any inference whose conclusion can be false, even though it has true premises. Thus, it will not do to dismiss induction by enumeration on grounds of Hume's argument and then accept some other mode of nondemonstrative inference without even considering how Hume's argument might apply to it. I am not arguing that Popper's method is incorrect.[35] I am not even arguing that Popper has failed in his attempt to justify this method. I do claim that Popper is engaged in the same task as many inductivists—namely, the task of providing some sort of justification for a mode of nondemonstrative inference. This enterprise, if successful, *is* a justification of induction.

4. Synthetic a priori Principles. A long philosophical tradition, dating back to antiquity, denies the empiricist claim that knowledge of the world rests solely upon observational evidence—that factual knowledge is limited to what we can observe and what we can infer therefrom. In the modern period, this rationalistic tradition is represented by men like Descartes and Leibniz who took their inspiration from the abstract aspect of modern physics. After Hume's devastating criticism of induction, Kant provided a more precise formulation, a fuller elaboration, and a more subtle defense of rationalism than any that had been given earlier (or, quite possibly, subsequently). As Kant himself testified, it was Hume who awakened him from his "dogmatic slumbers" and thereupon stimulated the *Critique of Pure Reason.*

The doctrine that there are synthetic a priori truths is, as I explained above, tantamount to the view that there are necessarily truth-preserving ampliative inferences. If we could find a *bona fide* demonstrative ampliative inference we would have a solution to Hume's problem of the ground of inference from the observed to the unobserved. This solution could be presented in either of two ways. First, one could assert that there are factual propositions that can be established by pure reason—without the aid of empirical evidence—and that these synthetic a priori propositions, in conjunction with premises established by observation, make it possible to deduce (nonampliatively) conclusions pertaining to unobserved matters of fact. Second, one could claim that these synthetic a priori propositions, although not added as premises to ampliative inferences to render them nonampliative, do instead provide a warrant for genuinely ampliative inferences from the observed to the unobserved. These alternatives are illustrated by inferences 2 and 1 on page 9. Inference 2 has been made out of 1 by the addition of a premise; 2 has been rendered nonampliative. If the premise added in 2 were synthetic a priori it would provide the ground for saying that 1, although ampliative, is necessarily truth-preserving and, hence, demonstrative. The synthetic a priori principle would constitute, in current parlance, an "inference ticket."

In order to appreciate the philosophical issues involved in the attempt to justify induction by means of a synthetic a priori principle, we must introduce some reasonably precise definitions of key terms. Two pairs of concepts are involved: first, the distinction between *analytic* and *synthetic* statements, and second, the distinction between *a priori* and *a posteriori* statements. I shall begin by discussing logical systems and defining some basic concepts relating to them. This will be a useful preliminary to the explications that are of primary concern.

A standard logical system contains clearly specified kinds of symbols that can be combined and manipulated according to certain explicit rules. The symbols will include logical constants such as "$\cdot$," "$\vee$," "$\sim$," "$\supset$," "(x)," and "$(\exists x)$" (which have, respectively, the following rough translations into English: "and," "or," "not," "if . . . then . . . ," "for all *x*," and "for some *x*"). Often the system includes variables for statements, "*p*," "*q*," "*r*,". . . . In addition, there are variables for individuals, "*x*," "*y*," "*z*,". . . , as well as predicate variables, "F," "G," "H,". . . , which stand for properties of individuals and relations among them. Some systems contain constants or proper names for individuals or predicates, but these are not necessary for our discussion. Formation rules specify the ways in which symbols may be combined; they define the concept of a well-formed-formula. For instance, "$(x)(Fx \vee Gx)$" is well-formed, while "$Fy \vee \supset (x)$" is not. The well-formed-formulas may be regarded as the meaningful formulas of the system in the sense that they are susceptible of meaningful interpretation. Formulas that are not well-formed are like nonsense strings of symbols. They are ungrammatical with regard to the rules of logical grammar in the way that "Edgar heaven unwise from without" is ungrammatical with regard to the rules of English grammar. Transformation rules, or rules of inference, provide means of deriving some formulas from others. They are the rules for manipulating formulas, and they define the concept of logical demonstration or proof.[36]

The formulas of a logical system can be interpreted by choosing some nonempty domain of individuals (concrete or abstract) and assigning meanings to the logical symbols with reference to the individuals of the chosen domain.[37] The logical constants are given their usual meanings, as indicated above; the statement connectives, for example, are interpreted in the usual truth table manner. The individual variables range over the individuals in the domain of interpretation, and the predicate variables refer to properties of and relations among these individuals. The truth values, truth and falsehood, are the values of the statement variables. Giving an interpretation consists in specifying the domain of individuals and assigning meanings to the symbols; the interpretation itself is the domain and the meanings assigned. When an interpretation is given, the well-formed-formulas of the logical system become true or false statements about the individuals of the domain. A logical system normally has many interpretations within any given domain of individuals, for there are various ways of assigning meanings to the nonlogical symbols.

Suppose, for instance, that we choose as the domain of interpretation the set of numbers {2, 4, 6, 8, 10}, and we let "F" stand for the property of being divisible by three. With this interpretation, the formula "$(x)Fx$," which means "every member of the domain is divisible by three," is false, while "$(\exists x)Fx$," which means "at least one member of the domain is divisible by three," is true. If, in a different interpretation within the same domain of individuals, we let "F" stand for the property of being even, the formula "$(x)Fx$," which now means "every member of the domain is even," becomes a true statement. On the other hand, if we choose a new domain consisting of all the integers from one to ten, retaining the same meaning "even" for "F," "$(x)Fx$" is rendered false.

We say that a given well-formed formula is *satisfied* in a given interpretation if it has become a true statement about the individuals of the domain as a result of that interpretation. We say that the formula is *satisfiable in that domain* if there is some interpretation in that domain in which it becomes a true statement. We say that a formula is *satisfiable* if there is some nonempty domain of individuals with respect to which the formula can be interpreted so as to become a true statement. A formula is *consistent* if and only if it is satisfiable; it is *inconsistent* if there is no interpretation in any nonempty domain within which it is satisfied. The denial of an inconsistent formula is a *valid* formula; a valid formula is one that is satisfied in every interpretation in every nonempty domain. A consistent formula whose denial is also consistent is *logically contingent.*

A valid formula is one that comes out true on every interpretation in every nonempty domain. While we know that it is impossible to have an axiomatic system of logic in which every valid formula is a provable theorem, we earnestly desire that only valid formulas shall be provable. If we find that a nonvalid formula can be deduced in a logical system, we modify or abandon the system. We shall say that any interpretation of a valid formula is a *logical truth.* A logical truth is any statement that results from any assignment of meanings to the symbols of a valid formula (provided this assignment does not violate the conditions demanded of all interpretations). For example,

$$(x)(\mathrm{F}x \supset \mathrm{G}x) \cdot (x)(\mathrm{G}x \supset \mathrm{H}x) \supset (x)(\mathrm{F}x \supset \mathrm{H}x)$$

$$\sim(p \vee q) \supset (\sim p \cdot \sim q)$$

are valid formulas; consequently, the following interpretations of them are logical truths:

If all cows are mammals and all mammals are warm-blooded, then all cows are warm-blooded.

If one can neither escape death nor escape taxes, then one cannot escape death and one cannot escape taxes.

This explication captures, I think, the intent of the traditional view that logical truths are propositions that hold in all possible worlds. Domains of interpretation play a role analogous to the notion of a possible world, and a valid formula is one that holds no matter how it is interpreted in any of these domains. A logical truth is any instance of a valid formula. Notice, however, that the definition of "valid formula" makes no reference to possible domains; it refers only to domains—i.e., actual domains. The reason that the qualification "possible" is not needed is that there are no impossible domains—to say that a domain is impossible would mean that it could not exist—so "impossible domains" are not available to be chosen as domains of interpretation.

Although it is reasonable to maintain, I think, that all logical truths are analytic, there seem to be analytic statements that are not logical truths. For instance,

All bachelors are unmarried

is not an interpretation of any valid formula. However, given the definition,

$$\text{Bachelor} =_{\text{df}} \text{unmarried adult male},$$

the foregoing statement can be transformed into a logical truth, for the definition gives license to substitute the *definiens*, "unmarried adult male," for the *definiendum* "bachelor." This substitution yields,

All unmarried adult males are unmarried,

which is an interpretation of the valid formula,

$$(x)(\mathrm{F}x \cdot \mathrm{G}x \cdot \mathrm{H}x \supset \mathrm{F}x) \,.$$

To incorporate cases of this sort, we may define an *analytic statement* as one that is a logical truth or can be transformed into a logical truth by definitional substitution of *definiens* for *definiendum*. The negation of an analytic truth is a *self-contradiction*. Any statement that is neither analytic nor self-contradictory is *synthetic*. More technically, we may define an *analytic statement* as one whose truth can be established solely by reference to the syntactic and semantic rules of the language, a *self-contradictory statement* as one whose falsity can be established solely by reference to the syntactic and semantic rules of the language, and a *synthetic statement* as one whose truth value is, in relation to the syntactic and semantic rules alone, indeterminate.[38]

Analytic statements, like logical truths, have been characterized as statements that are true in all possible worlds. I have already explained how I think this characterization applies to logical truths. When the class of analytic statements is constructed by augmenting logical truths with statements that can be reduced to logical truths by definitional substitution, it is evident that these additional statements are true in all possible worlds in just the same manner as are logical truths. A definitional substitution cannot, after all, alter the truth value of any statement.

Analytic statements have sometimes been characterized as statements that are true by definition.[39] The general idea behind this formulation is sound in the sense I have been trying to explicate, but it is also misleading. We do not give a direct definition of a sentence as true; rather, when we provide definitions and other linguistic conventions for our language, certain statements have to be true in consequence. Analytic statements have often been characterized as statements whose truth depends upon the definitions of the terms occurring in them. Again, the idea is fundamentally correct, but the formulation is faulty. In a certain trivial sense, the truth of any statement depends upon the definitions of the terms in it. On a cloudy day, the statement, "The sky is blue," is false, but it can be transformed into a true statement if we are willing to redefine the word "blue" so that it means what we usually mean by "gray." On either meaning, however, the truth or falsity of the statement depends not only upon the definition of "blue" and the other words in it, but also upon some nonlinguistic meteorological facts. Every statement depends for its truth or falsity partly upon linguistic conventions; the truth values of analytic and self-contradictory statements depend entirely upon linguistic considerations.[40]

Analytic statements are often said, moreover, to be devoid of any factual content. Although there are difficulties in giving an adequate account of the concept of factual content, enough can be said to illuminate its relation to analytic statements. The basic feature seems to be that factual content of a statement is a measure of the capacity of that statement to *rule out* possibilities. In this respect, it is a negative concept. In a state of total ignorance all possible states of affairs are live possibilities; any possible state of affairs might, for all we know, be the actual state of things. As knowledge accumulates, we realize that some of the possibilities are not actualized. The statements expressing our knowledge are incompatible with descriptions of various possible worlds, so we know that these possibilities are ruled out—our actual world does not coincide with any of these possibilities that are incompatible with what we know. Generally speaking, moreover, the greater our knowledge —the greater the factual content of the statements we know—the more possibilities are disqualified from being actual. Imagine, for instance, the inhabitants of Plato's famous cave, who are totally ignorant of the nature of the external world. They can imagine birds of all sorts, including ravens of various colors. When the emissary to the outer world returns and reports that all ravens are black, those who remained in the cave can rule out all possibilities that had room for ravens of other colors. The statement, "All ravens are black," has factual content because of the descriptions of possible worlds with which it is incompatible. If, however, the emissary should return and remark that every raven is either black or nonblack, his statement would be totally lacking in content, and the permanent inhabitants of the cave—anxious for knowledge of the external world—would be justly furious with him for his empty report. His statement would lack content because it is compatible with every possibility. It is an interpretation of a valid formula, so it is a logical truth. It is an interpretation of a formula that cannot have a false interpretation—a formula that is true under any interpretation in any nonempty domain. Since it is true under any possible circumstances and is not incompatible with any description of a possible world, its content is zero. Any analytic statement will, as we have seen above, share this characteristic. We are, therefore, entitled to assert that analytic statements have no factual content. Synthetic statements, on the other hand, are interpretations of logically contingent formulas. Since the denial of a logically contingent formula has true interpretations in nonempty domains, knowledge that a synthetic statement holds does rule out some possibilities, so the synthetic statement does have factual content.

It is worth noting that the nonampliative nature of valid deduction is a consequence of the foregoing characterization of factual content. Take any valid deduction from premises $P_1, P_2, \ldots, P_n$ and conclusion C. The conditional statement, $P_1 \cdot P_2 \cdot \ldots \cdot P_n \supset C$, formed by taking the conjunction of the premises of the deduction as antecedent and the conclusion of the deduction as consequent, is analytic. This conditional statement, therefore, holds in all possible worlds; consequently, the set of possible worlds in which the antecedent is true is a subset of the set of possible worlds in which the consequent is true. It follows that the set of all possible worlds *excluded* by the conclusion is contained within the set of all possible worlds *excluded* by the premises. The factual content of the premises is, therefore, at least as great as that of the conclusion, and the factual content of the premises includes the factual content of the conclusion. Moreover, the content of the premises is equal to the combined factual content of all valid deductive conclusions of these premises, for any set of premises follows deductively from the conjunction of all conclusions that can be validly deduced from them. These considerations establish the claim that valid deductions are nonampliative, and the claim that the deductive consequences of a statement reveal its factual content.

Let us now turn to the other distinction required for our discussion. A statement is a priori if its truth or falsity can be established without recourse to observational evidence; it is a posteriori if observational evidence is needed to establish its truth or falsity. The distinction between a priori and a posteriori statements refers exclusively to the justification of statements and has nothing to do with discovery. The statements of arithmetic, for example, are regarded by most philosophers as a priori; the fact that children may learn arithmetic by counting physical objects (e.g., fingers) has nothing to do with the issue. Arithmetic statements can be established formally, without the aid of empirical observation or experiment, and this qualifies them as a priori. It is evident, moreover, that analytic statements, as they have been described above, are a priori. Since their truth follows from logical truths and definitions alone—that is, from syntactical and semantical considerations alone—observation and experiment are not required for their proof.

Most philosophers would acknowledge that many synthetic statements are a posteriori. It would seem that no amount of pure ratiocination would reveal whether I had eggs for breakfast this morning or whether there is a typewriter on the desk in the next office. Some sort of observation would seem to be indispensable. However, it is not nearly as evident that *all* synthetic statements are a posteriori. The doctrine that there are synthetic a priori statements is, I take it, the thesis of rationalism. It was maintained by Kant, as well as by many other philosophers both before and after him. The doctrine that all a priori statements are either analytic or self-contradictory is the thesis of empiricism as I understand it.

I know of no easy way to argue the question of whether there are any synthetic a priori statements. The history of human thought has provided many attempts to establish synthetic a priori truths, with a notable lack of success in my opinion. It is interesting to note some of the more important statements that have been alleged to enjoy this status. From antiquity, the proposition that nothing can be created out of nothing—*ex nihilo nihil fit*—has been taken as a self-evident truth of reason. It was accepted by the ancient atomist Lucretius,[41] although it is denied by contemporary steady-state cosmologists who are prepared to maintain that hydrogen atoms are created *ex nihilo*.[42] These cosmologists are sometimes ridiculed by cosmologists of other persuasions because of their rejection of the ancient principle. A closely related principle that has a long history as a putative synthetic a priori truth is the principle of sufficient reason. Although formulated in a variety of ways, it says roughly that nothing happens unless there is a sufficient reason for it. Lucretius, however, rejected the principle of sufficient reason, for he maintained that his atoms swerved *for no reason of any kind*.[43] He has often been disparaged for this doctrine by those who maintain the principle of sufficient reason, and typical college sophomores today still find the concept of a genuinely undetermined event unintelligible. There *must* be *some* reason for the swerving, they say. At the same time, modern physical theory appears to proceed in violation of the principle of sufficient reason. Quantum theory denies, for example, that there is a sufficient reason for the radioactive decay of one particular atom rather than another at some particular time. Quantum theory may be false, but if so its falsity must be shown empirically. Attempts to provide a priori demonstrations of the falsity of indeterministic quantum mechanics are, I believe, entirely unsuccessful.

Descartes provides an especially clear example of the use of synthetic a priori principles to justify ampliative inference. Starting from his famous *cogito,* he set out to deduce a complete account of the real world. He never supposed that nonampliative deduction would be equal to the task; instead, he appealed to principles he considered evident to the natural light of reason: "Now it is manifest by the natural light that there must at least be as much reality in the efficient and total cause as in its effect. For, pray, whence can the effect derive its reality, if not from its cause?" [44] The man who thought he could not be certain that $2 + 2 = 4$ or that he had hands unless he could prove that God is not a deceiver found the foregoing principle so clear and distinct that it is impossible to conceive its falsity!

Contemporary philosophy has not been lacking in claims for the synthetic a priori status of a variety of statements, although frequently they seem to be far less fancy than those for which the traditional claim was made. A recent important text in the philosophy of science argues that the statement,

No event [temporally] precedes itself

qualifies for the position.[45] However, the author fails to consider the empirical hypothesis that the history of the universe is a closed cycle. If this were true, events would precede themselves.[46] Since the statement in question is incompatible with an empirical hypothesis, it could be falsified empirically, and this means that it cannot be an a priori truth. Another currently famous example is the statement,

Anything colored is extended.[47]

Again, it seems to me, the statement turns out on close examination to be a posteriori. It is a physical fact, established empirically, that the stimulus that normally gives rise to color sensations is a very short wave. We find, by experience, that short waves provide rather reliable indications of the surface characteristics of objects from which they are emitted or reflected. If our visual sensations were elicited by long waves, like radio waves, they would be found not to provide much useful information about surfaces from which they emanate; in particular, they would be rather uninformative about size. Under these circumstances, I seriously doubt that we would closely associate the qualities of color and extensiveness, for we do not as a matter of fact make a close association between auditory qualities and extension. These considerations provide grounds, I believe, for denying that the foregoing statement is actually synthetic a priori.

Although Kant's attempt to supply a synthetic a priori grounding for ampliative inference was by no means the first nor the last, it is, in many respects, the most notable. According to his view, the statements of geometry and arithmetic are synthetic a priori. The case of geometric propositition is clearest.[48]

Since Euclid at least, geometry had been an a priori science, and until the seventeenth century it was the only well-developed science we had. There was, consequently, a tradition of more than two millenia in which geometry was *the* science—the model for all scientific knowledge. Geometry had existed throughout all this time in an axiomatic form; the theorems of the system were thought to be purely deductive consequences of a small number of postulates. Moreover, there was one unique set of postulates. It was not until the discovery of non-Euclidean geometry by Bolyai and Lobachewsky early in the nineteenth century that anyone thought very seriously about alternatives. Until the advent of non-Euclidean geometry, the postulates of Euclidean geometry were regarded as self-evident truths—as propositions whose denials could not be reasonably entertained. Thus, until the early part of the nineteenth century at least, the a priori status of geometry seemed completely secure.[49]

Long before the discovery of non-Euclidean geometries it was widely recognized that experience with physical objects has substantial heuristic value. As we all know, the Egyptians discovered a number of important geometric truths in dealing with practical problems of surveying and engineering. Even Plato—a rationalist of the first order—recognized that experience in dealing with physical objects that approximate ideal geometrical figures is an indispensable part of the education of a geometer. But these heuristic considerations have nothing to do with the question of whether geometry is a priori or a posteriori. Empirical observations fulfill much the same function as the diagrams in the geometry text; they help the student to understand the subject, but they have no logical place in the proofs as such.

The synthetic status of geometry also seemed well assured during the long reign of Euclidean geometry. The postulates and theorems appeared to have real factual content; they seemed to provide indispensable aid in making ampliative inferences regarding the spatial characteristics of actual physical objects. Consider an example. A farmer has a flat rectangular field with one side 90 yards long and another side 120 yards long. He wants to divide the field diagonally with a wire fence, so using the Pythagorean Theorem he calculates the required length of fence wire to be 150 yards. He goes to town and buys that much wire, and finds that it does, indeed, just reach across the diagonal distance he wants to fence. There is no doubt that he has made an ampliative inference, for there is no logical contradiction in supposing that the wire would be too long or too short. He has made a successful ampliative inference from the observed to the unobserved; because of the propositions of geometry it was a necessary inference as well. Thus, so it seemed, geometry did embody truly synthetic a priori propositions.

Non-Euclidean geometry was discovered a bare twenty years after Kant's death, and it was later proved that Euclidean and non-Euclidean geometries are equally consistent. If non-Euclidean geometry contains a contradiction, so does Euclidean geometry, and if Euclidean geometry contains a contradiction, so does non-Euclidean geometry. All these geometries are on a par from the standpoint of logical consistency. This is sometimes taken to be a refutation of Kant's doctrine of the synthetic a priori character of geometry, but to do so is a mistake. To be sure, the existence of non-Euclidean geometries provides an impetus for the careful examination of the Kantian thesis, but it does not by itself disprove it. Kant himself entertained the possibility of logically consistent alternatives to Euclidean geometry. After all, Kant was *not* trying to show that geometrical propositions are logically necessary; if they were, they would be analytic, not synthetic. Kant was attempting to establish a different sort of necessity for them.

According to Kant, geometry constitutes a necessary form into which our experience of objective physical reality must fit. It is a necessary form of visualization. Kant admitted the possibility of alternative types of geometry as formal toys for mathematicians to play with, but they were epistemologically unsuitable as forms for experience of the objective world. Kant claims, therefore, that Euclidean geometry, although equal to the non-Euclidean geometries from the purely logical aspect, held a privileged position from an *epistemological* standpoint. This view seems widely held today, for one often hears remarks to the effect that it is impossible to visualize non-Euclidean geometries and that Euclidean geometry is the only one that can be visualized. Kant held it to be a necessary characteristic of the human mind that it must organize sense experiences according to the forms established by Euclidean geometry.

Subsequent philosophical analysis, primarily by Helmholtz and Reichenbach,[50] has shown two things, I believe. First, the alleged inability of people to visualize non-Euclidean geometries, if such inability does obtain, is a fact of empirical psychology; it is not a necessary aspect of the human intellect that can be demonstrated philosophically. Second, the supposition that we cannot visualize non-Euclidean geometries is probably false. This false supposition is based partly upon the lack of clarity about what we mean by "visualize" and partly upon the accident of being born into a world in which Euclidean relations are frequently exemplified by physical objects (at least to a high degree of approximation) and non-Euclidean relations are not.

The net outcome of the analysis of geometry is a rejection of the doctrine that geometry is synthetic a priori. To see this, it is useful to distinguish pure and applied geometry. Pure geometry is a strictly formal discipline concerned solely with what theorems follow deductively from what postulates. No attempt is made to elevate any particular set of postulates to a privileged position. *Pure geometry is a priori, but it is* not *synthetic.* Within the domain of pure geometry as such, there is no room for the question of which of the various geometries actually describes the spatial relations among real physical objects. To raise this question we must move to the domain of applied geometry. This question is empirical. From the logical standpoint of consistency, all the geometries are on a par. From the epistemological standpoint of visualizability, they are also on a par. The choice of a geometry to describe the world depends upon the behavior of physical objects, and that can be determined only by observation. Thus, *applied geometry is synthetic, but it is* not *a priori.*

Kant maintained that arithmetic, as well as geometry, embodies synthetic a priori truths. In their monumental work, *Principia Mathematica,* Russell and Whitehead attempt to refute this doctrine by showing that arithmetic can be reduced to logic; from this it would follow that the statements of arithmetic, although a priori, are analytic rather than synthetic. Full agreement has not yet been reached regarding the final success of the Russell-Whitehead program, but nothing emerges from it to give any aid or comfort to the defenders of the synthetic a priori in arithmetic. A more satisfactory account will, perhaps, distinguish pure from applied arithmetic just as we distinguish pure from applied geometry. In that case, the result would parallel that of geometry: Pure arithmetic is a priori but not synthetic, while applied arithmetic is synthetic but not a priori.[51]

Kant's approach to the question of synthetic a priori principles is profoundly instructive. So convinced was he that geometry provided examples of synthetic a priori propositions that he did not need to tarry long over the question of whether there are any such things. Instead, he moved on the question of how they are possible. Synthetic a priori knowledge (if there is such) does exhibit a genuine epistemological mystery. After some exposure to formal logic one can see without much difficulty how linguistic stipulations can yield analytic statements that hold in any possible world. It is easy to see that "Snow is white or snow is not white" is true simply because of the meanings we attach to "or" and "not." Analytic a priori statements are no great mystery. Likewise, it is not too difficult to see how our senses can provide clues to the nature of physical reality, helping us to establish propositions that are true in some but not all possible worlds. Although there are epistemological problems of perception, like illusion, that must not be ignored, we can still understand in very general terms how there can be a posteriori knowledge of synthetic propositions. But how could we conceivably establish by pure thought that some logically consistent picture of the real world is false? How could we, without any aid of experience whatever, find out anything about our world in contradistinction to other possible worlds? Given a logically contingent formula—one that admits of true as well as false interpretations—how could we hope to decide on a completely a priori basis which of its interpretations are true and which false? The empiricist says it is impossible to do so, and in this I think he is correct. Nevertheless, it is tempting to endow various principles with the status of synthetic a priori truths. It was to Kant's great credit that he saw the urgency of the question: *How is this possible?*

Various causal principles, as we have seen, have been accorded the status of synthetic a priori truths—for example, the traditional *ex nihilo* principle, the principle of sufficient reason, and Descartes' principle that the cause must be as great as the effect. Kant also, in addition to claiming that the propositions of arithmetic and geometry are synthetic a priori, maintained that the principle of universal causation—everything that happens presupposes something from which it follows according to a rule —is synthetic a priori.[52] It is by means of this principle that he hoped to dispose of the problem of induction. However, Kant's attempt to explain the possibility of synthetic a priori propositions is unsatisfactory. The propositions of Euclidean geometry do not enjoy epistemological primacy; the propositions of arithmetic lack synthetic content, and the physical world can be made intelligible in nondeterministic terms. Human powers of conception and visualization far exceed the limits Kant saw as necessary constraints upon the human intellect and as the source of synthetic a priori truths.

It is unfortunate that subsequent philosophers have paid little attention to Kant's central question: How are synthetic a priori propositions possible? Instead, the category of synthetic a priori propositions has too often become a convenient wastebasket for statements not readily classifiable as analytic or a posteriori. The contents of this wastebasket may, of course, turn out to be very handy in dealing with tough philosophical problems, but the crucial point is whether the wastebasket is really empty. It seems to me that all such statements can be shown, on careful examination, to be analytic or a posteriori, and that no convincing example of a synthetic a priori proposition has yet been produced. Even if this is so, of course, it does not prove that there are no synthetic a priori statements. It should, however, give us pause, and it does relieve us of any obligation to accept the positive rationalistic thesis that there are synthetic a priori propositions. It does place the burden of proof upon those who hope to escape Hume's problem of induction by way of a synthetic a priori principle. Moreover, even if a recalcitrant example were given—one that seemed to defy all analysis as either analytic or a posteriori—it might still be reasonable to suppose that we had not exercised sufficient penetration in dealing with it. If we are left with a total epistemological mystery on the question of how synthetic a priori propositions are possible, it might be wise to suppose it more likely that our analytic acumen is deficient than that an epistemological miracle has occurred.

5. *The Principle of Uniformity of Nature.* A substantial part of Hume's critique of induction rested upon his attack on the principle of the uniformity of nature. He argued definitively that the customary forms of inductive inference cannot be expected to yield correct predictions if nature fails to be uniform—if the future is not like the past—if like sensible qualities are not accompanied by like results.

> All inferences from experience suppose, as their foundation, that the future will resemble the past, and that similar powers will be conjoined with similar sensible qualities. If there be any suspicion that the course of nature may change, and that the past may be no rule for the future, all experience becomes useless, and can give rise to no inference or conclusion.[53]

He argued, moreover, that there is no logical contradiction in the supposition that nature is not uniform—that the regularities we have observed up to the present will fail in wholesale fashion in the future.

> It implies no contradiction that the course of nature may change, and that an object, seemingly like those which we have experienced, may be attended

with different or contrary effects. May I not clearly and distinctly conceive that a body, falling from the clouds, and which, in all other respects, resembles snow, has yet the taste of salt or feeling of fire? Is there any more intelligible proposition than to affirm, that all the trees will flourish in December and January, and decay in May and June? Now whatever is intelligible, and can be distinctly conceived, implies no contradiction, and can never be proved false by any demonstrative argument . . .[54]

He argues, in addition, that the principle of uniformity of nature cannot be established by an inference from experience: "It is impossible, therefore, that any arguments from experience can prove this resemblance of the past to the future; since all these arguments are founded on the supposition of that resemblance."[55] Throughout Hume's discussion there is, however, a strong suggestion that we might have full confidence in the customary inductive methods if nature were known to be uniform.

Kant attempted to deal with the problem of induction in just this way, by establishing a principle of uniformity of nature, in the form of the principle of universal causation, as a synthetic a priori truth. Kant claimed, in other words, that every occurrence is governed by causal regularities, and this general characteristic of the universe can be established by pure reason, without the aid of any empirical evidence. He did not try to show that the principle of universal causation is a principle of logic, for to do so would have been to show that it was analytic—not synthetic—and thus lacking in factual content. He did not reject Hume's claim that there is no logical contradiction in the statement that nature is not uniform; he did not try to prove his principle of universal causation by deducing a contradiction from its denial. He did believe, however, that this principle, while not a proposition of pure logic, is necessarily true nevertheless. Hume, of course, argued against this alternative as well. He maintained not only that the uniformity of nature is not a logical or analytic truth, but also that it cannot be any other kind of a priori truth either. Even before Kant had enunciated the doctrine of synthetic a priori principles, Hume had offered strong arguments against them:

I shall venture to affirm, as a general proposition, which admits of no exception, that the knowledge of this relation [of cause and effect] is not, in any instance, attained by reasonings *a priori*.[56]

Adam, though his rational faculties be supposed, at the very first, entirely perfect, could not have inferred from the fluidity and transparency of water that it would suffocate him, or from the light and warmth of fire that it would consume him.[57]

When we reason *a priori*, and consider merely any object or cause, as it appears to the mind, independent of all observation, it never could suggest to us the notion of any distinct object, such as its effect; much less, show us the inseparable and inviolable connexion between them. A man must be very sagacious who could discover by reasoning that crystal is the effect of heat, and ice of cold, without being previously acquainted with the operation of these qualities.[58]

Now whatever is intelligible, and can be distinctly conceived . . . can never be proved false by any . . . abstract reasoning *a priori.*[59]

Hume argues, by persuasive example and general principle, that nothing about the causal structure of reality can be established by pure reason. He poses an incisive challenge to those who would claim the ability to establish a priori knowledge of a particular causal relation or of the principle of universal causation. In the foregoing discussion of synthetic a priori statements, I have given reasons for believing that Kant failed to overcome Hume's previous objections.

There is, however, another interesting issue that arises in connection with the principle of uniformity of nature. Suppose it could be established—never mind how—prior to a justification of induction. Would it then provide an adequate basis for a justification of induction? The answer is, I think, negative.[60]

Even if nature is uniform to some extent, it is not absolutely uniform. The future is something like the past, but it is somewhat different as well. Total and complete uniformity would mean that the state of the universe at any given moment is the same as its state at any other moment. Such a universe would be a changeless, Parmenidean world. Change obviously does occur, so the future is not exactly like the past. There are some uniformities, it appears, but not a complete absence of change. The problem is how to ferret out the genuine uniformities. As a matter of actual fact, there are many uniformities *within experience* that we take to be mere coincidences, and there are others that seem to represent genuine causal regularities. For instance, in every election someone finds a precinct, say in Maryland, which has always voted in favor of the winning presidential candidate. Given enough precincts, one expects this sort of thing by sheer chance, and we classify such regularities as mere coincidences. By contrast, the fact that glass windowpanes break when bricks are hurled at them is more than mere coincidence. Causal regularities provide a foundation for inference from the observed to the unobserved; coincidences do not. We can predict with some confidence that the next glass window pane at which a brick is hurled will break; we take with a grain of salt the prediction of the outcome of a presidential election early on election night when returns from the above-mentioned precinct are in. The most that a principle of uniformity of nature could say is that there are some uniformities that persist into the future; if it stated that every regularity observed to hold within the scope of our experience also holds universally, it would be patently false. We are left with the problem of finding a sound basis for distinguishing between mere coincidence and genuine causal regularity.

Kant's principle of universal causation makes a rather weak and guarded statement. It asserts only that there exist causal regularities: "Everything that happens presupposes something from which it follows according to some rule." For each occurrence it claims only the existence of *some* prior cause and *some* causal regularity. It gives no hint as to how we are to find the prior cause or how we are to identify the causal regularity. It therefore provides no basis upon which to determine whether the inductive inferences we make are correct or incorrect. It would be entirely consistent with Kant's principle for us always to generalize on the basis of observed coincidences and always to fail to generalize on the basis of actual causal relations. It would be entirely consistent with Kant's principle, moreover, for us always to cite a coincidentally preceding event as the cause instead of the event that is the genuine cause. Kant's principle, even if it could be established, would not help us to justify the assertion that our inductive inferences would always or usually be correct. It would provide no criterion to distinguish sound from unsound inductions. Even if Kant's program had succeeded in establishing a synthetic a priori principle of universal causation, it would have failed to produce a justification of induction.

6. The Postulational Approach. Some philosophers, chary of synthetic a priori principles and pessimistic about the possibility of overcoming Hume's skeptical arguments concerning induction, have turned to the elaboration of postulates for scientific inference. They have sought, sometimes with great skill and ingenuity, for the kinds of assumptions which, if accepted, would render scientific method acceptable. They have assiduously tried to avoid both extremes observed in the preceding section: principles too strong to be true or too weak to be useful.

One notable recent attempt to provide a postulational foundation for scientific inference is due to Bertrand Russell.[61] After analyzing and rejecting J. M. Keynes' famous "postulate of limited independent variety,"[62] Russell goes on to propound a relatively complex set of five postulates of his own. He maintains that without some synthetic assumptions about the world it is impossible to infer from the observed to the unobserved. The alternative to adopting postulates of scientific inference is solipsism of the moment. "If anyone chooses to maintain solipsism of the moment, I shall admit that he cannot be refuted, but shall be profoundly skeptical of his sincerity."[63] The only tenable approach is to find assumptions that are "logically necessary if any occurrence or set of occurrences is ever to afford evidence in favor of any other occurrence."[64]

Owing to the world being such as it is, certain occurrences are sometimes, in fact, evidence for certain others; and owing to animals being adapted to their environment, occurrences which are, in fact, evidence of others tend to arouse expectation of those others. By reflecting on this process and refining it, we arrive at the canons of inductive inference. These canons are valid if the world has certain characteristics which we all believe it to have.[65]

Russell offers the following five postulates to serve the intended function:[66]

1. The postulate of quasi-permanence: "*Given any event A, it happens very frequently that, at any neighboring time, there is at some neighboring place an event very similar to A.*"
2. The postulate of separable causal lines: "*It is frequently possible to form a series of events such that from one or two members of the series something can be inferred as to all the other members.*"
3. The postulate of spatio-temporal continuity in causal lines: "This postulate is concerned to deny 'action at a distance,' and to assert that when there is a causal connection between two events that are not contiguous, there must be intermediate links in the causal chain such that each is contiguous to the next."
4. The structural postulate: "*When a number of structurally similar complex events are ranged about a center in regions not widely separated, it is usually the case that all belong to causal lines having their origin in an event of the same structure at the center.*"
5. The postulate of analogy: "*Given two classes of events A and B, and given that, whenever both A and B can be observed, there is reason to believe that A causes B, then if, in a given case, A is observed, but there is no way of observing whether B occurs or not, it is probable that B occurs; and similarly if B is observed, but the presence or absence of A cannot be observed.*"

Russell discusses in detail the function he claims for these postulates in inference from the observed to the unobserved. Each of the postulates asserts that something happens frequently, but not invariably, so the postulates support inferences whose conclusions can be accepted with a degree of confidence but not with certainty. This does not mean, of course, that such inferences are useless. There is no reason to regard the fallibility of scientific inference as a devastating criticism of it.

While Russell argues that some synthetic postulates are *necessary* for scientific inference, he does not wish to maintain that those he has stated enjoy that status.

> The above postulates are probably not stated in their logically simplest form, and it it likely that further investigation would show that they are not all necessary for scientific inference. I hope and believe, however, that they are sufficient. . . .
>
> The postulates, in the form in which I have enunciated them, are intended to justify the first steps toward science, and as much of common sense as can be justified. My main problem in this Part has been epistemological: What must we be supposed to know, in addition to particular observed facts, if scientific inferences are to be valid?[67]

The first task that demands attention, if we are to evaluate this postulational approach, is to straighten out the relations of implication that are being asserted. There is considerable vagueness on this score. For one thing, there seems to be a certain amount of vacillation between the search for necessary conditions and for sufficient conditions, and for another, Russell is not unequivocal regarding the question of what the condition, be it necessary or sufficient, is to be a *condition of*. Consider the second point first. As the foregoing quotations reveal, Russell sometimes speaks as if he is seeking a condition for the possibility of any kind of inference whatsoever from one matter of fact to another—for the possibility that any fact could constitute evidence of any sort for any other fact. At other times he appears to be looking for a condition for the general correctness of the sorts of inference from the observed to the unobserved we usually regard as correct within science and common sense. Enough has been said already to show that these two desiderata are by no means identical, for we have made reference above to a variety of alternative methods: induction by enumeration, crystal gazing, the hypothetico-deductive method, the counterinductive method, and the method of corroboration. We shall see below that there is no trouble at all in describing an infinite array of distinct, conflicting methods of ampliative inference. To say merely that ampliative inference is possible would mean only that some of these methods are correct; to say merely that one occurrence can constitute evidence for another is to say the same. Neither of these claims goes far toward handling the problem of induction, which is, basically, the problem of choosing the appropriate method or methods from among the infinity of available candidates. Moreover, the whole spirit of Russell's discussion indicates that he is assuming the soundness by and large of scientific knowledge as presently constituted. This is, of course, much stronger than the assumption that knowledge is possible at all. To assume the fundamental correctness of the methods of empirical science is, obviously, to assume an answer to Hume's problem of the justification of induction. To make such an assumption is not to provide a solution of the problem.

Granting for the moment, however, that we are looking for a condition of the validity of scientific inference as generally practiced, is it a necessary or a sufficient condition we seek? One common way of putting the question is to ask: What are the *presuppositions* of induction or scientific inference? To answer this question we must first be clear about the meaning of the term "presuppose." Consider a simple and typical example: Receiving a scholarship *presupposes* being admitted as a student; likewise, being a student *is a presupposition of* the receipt of a scholarship. In the foregoing statement, the term "presupposes" can be translated "implies." Receiving a scholarship implies being a student, but being a student does not imply receiving a scholarship, for many students pay their tuition. In general, "*A* presupposes *B*" can be translated "*A* implies *B*," which means, in turn, "*A* is a *sufficient condition* of *B*." Moreover, "*A* presupposes *B*" means "*B* is presupposed by *A*" or "*B* is a presupposition of *A*." Thus, a presupposition of *A* is a *necessary condition* of *A*, that is, it is anything that *A* implies. A presupposition of induction would be any statement *implied by* the assertion, "Induction is a correct method of inference from the observed to the unobserved." A presupposition of the validity of scientific inference would be any statement *implied by* the assertion, "Scientific knowledge is, by and large, well founded." We cannot conclude from any presupposition of induction or scientific method that such methods are correct, for the implication goes in the opposite direction. To justify scientific inference we need a postulate that is a *sufficient condition* of the correctness of induction. This is, after all, what Russell claims for his postulates. But not just any sufficient condition will do; the search is for sufficient conditions that are both minimal and plausible. We should not assume more than necessary, and we should not assume postulates that are obviously false. It appears, then, that Russell is seeking a condition that is both necessary and sufficient to the suitability of scientific modes of inference—that is, a statement to fill the blank in "Induction is an acceptable inferential tool if and only if . . ." "Induction is an acceptable inferential tool" obviously works, but we hope to find a more interesting statement.

When we realize that there are many alternative modes of ampliative inference, we are naturally led to ask in what sorts of circumstances different ones will work better than their fellows. There is no way, as we have learned from Hume, to prove that any one of them will work successfully in all conceivable circumstances, so we might reasonably attempt to characterize the kinds of universes in which each is most suitable. Thus, instead of looking for postulates to justify *the* inductive method, we seek different sets of postulates, each set being sufficient to justify a different inductive method. Since the alternative inductive methods are in mutual conflict, we expect to find sets of postulates that are mutually incompatible. Arthur Burks has pursued this line of inquiry in some detail, and he has exhibited, by way of example, three radically different and conflicting inductive methods, each of which would require a distinct set of postulates.[68] If we could find some ground for holding one such postulate set in preference to all competing sets, that would constitute a justification of induction, for we would have justified acceptance of those postulates and *ipso facto* of the consequence that a certain method works. Such a justification is, of course, precisely the sort that Hume's arguments rule out.

Russell writes as if we have a simple choice between accepting his postulates (or some suitable modification thereof) and embracing solipsism of the moment. The situation is not that simple. We have a choice between accepting Russell's postulates and a wide variety of other conflicting postulates. We cannot pretend to know, except by inductive reasoning, which ones are true. We cannot use inductive inference to establish one set of postulates in preference to the others on pain of circularity. The most we can hope to establish is a series of conditional statements of the form, "If postulate set P holds, then inductive method M will work (at least decently often)." We cannot hope to show the unconditional utility of any method. Such a result can hardly be said to do justice to the method of science. In astronomy we predict an eclipse unconditionally. We do not say, either explicitly or implicitly, "If Russell's five postulates hold, then the sun's disc will be obscured by the moon at a specified time from a particular vantage point." From the postulational standpoint, however, the most we can assert is the conditional. Science, as a result, would be empty.

If a philosopher embraces the postulational approach to induction, he must not boggle at frankly making factual assumptions without attempting any justification of them. This is clearly an admission of defeat regarding Hume's problem, but it may be an interesting way to give up on the problem. The search for the weakest and most plausible assumptions sufficient to justify various alternative inductive methods may cast considerable light upon the logical structure of scientific inference. But, it seems to me, admission of unjustified and unjustifiable postulates to deal with the problem is tantamount to making scientific method a matter of faith. I shall have more to say on that subject while discussing the significance of the problem of induction.[69]

7. A Probabilistic Approach. It may seem strange in the extreme that this discussion of the problem of induction has proceeded at such great length without seriously bringing in the concept of probability. It is very tempting to react immediately to Hume's argument with the admission that we do not have *knowledge* of the unobserved. Scientific results are not established with absolute certainty. At best we can make probabilistic statements about unobserved matters of fact, and at best we can claim that scientific generalizations and theories are highly confirmed. We who live in an age of scientific empiricism can accept with perfect equanimity the fact that the quest for certainty is futile; indeed, our thanks go to Hume for helping to destroy false hopes for certainty in science.

Hume's search for a justification of induction, it might be continued, was fundamentally misconceived. He tried to find a way of proving that inductive inferences with true premises would have *true* conclusions. He properly failed to find any such justification precisely because it is the function of *deduction* to prove the truth of conclusions, given true premises. Induction has a different function. An inductive inference with true premises establishes its conclusions as *probable*. No wonder Hume failed to find a justification of induction. He was trying to make induction into deduction, and he succeeded only in proving the platitude that induction is not deduction.[70] If we want to justify induction, we must show that inductive inferences establish their conclusions as probable, not as true.

The foregoing sort of criticism of Hume's arguments is extremely appealing, and it has given rise to the most popular sort of attempt, currently, to deal with the problem.[71] In order to examine this approach, we must consider, at least superficially, the meaning of the concept of probability. Two basic meanings must be taken into account at present.

One leading probability concept identifies probability with frequency —roughly, the probable is that which happens often, and the improbable is that which happens seldom. Let us see what becomes of Hume's argument under this interpretation of probability. If we were to claim that inductive conclusions are probable in this sense, we would be claiming that inductive inferences with true premises often have true conclusions, although not always. Hume's argument shows, unhappily, that this claim cannot be substantiated. It was recognized long before Hume that inductive inferences cannot be expected always to lead to the truth. Hume's argument shows, not only that we cannot justify the claim that *every* inductive inference with true premises will have a true conclusion, but also, that we cannot justify the claim that *any* inductive inference with true premises will have a true conclusion. Hume's argument shows that, for all we can know, every inductive inference made from now on might have a false conclusion despite true premises. Thus, Hume has proved, we can show neither that inductive inferences establish their conclusions as true nor that they establish their conclusions as probable in the frequency sense. The introduction of the frequency concept of probability gives no help whatever in circumventing the problem of induction, but this is no surprise, for we should not have expected it to be suitable for this purpose.

A more promising probability concept identifies probability with degree of rational belief. To say that a statement is probable in this sense means that one would be rationally justified in believing it; the degree of probability is the degree of assent a person would be rationally justified in giving. We are not, of course, referring to the degree to which anyone *actually* believes in the statement, but rather to the degree to which one could *rationally* believe it. Degree of actual belief is a purely psychological concept, but degree of rational belief is determined objectively by the evidence. To say that a statement is probable in this sense means that it is supported by evidence. But, so the argument goes, if a statement is the conclusion of an inductive inference with true premises, it *is* supported by evidence—by inductive evidence—this is part of what it *means* to be supported by evidence. The very concept of evidence depends upon the nature of induction, and it becomes incoherent if we try to divorce the two. Trivially, then, the conclusion of an inductive inference is probable under this concept of probability. To ask, with Hume, if we should accept inductive conclusions is tantamount to asking if we should fashion our beliefs in terms of the evidence, and this, in turn, is tantamount to asking whether we should be rational. In this way we arrive at an "ordinary language dissolution" of the problem of induction. Once we understand clearly the meanings of such key terms as "rational," "probable," and "evidence," we see that the problem arose out of linguistic confusion and evaporates into the question of whether it is rational to be rational. Such tautological questions, if meaningful at all, demand affirmative answers.

Unfortunately, the dissolution is not satisfactory.[72] Its inadequacy can be exhibited by focusing upon the concept of inductive evidence and seeing how it figures in the foregoing argument. The fundamental difficulty arises from the fact that the very notion of inductive evidence is determined by the rules of inductive inference. If a conclusion is to be supported by inductive evidence, it must be the conclusion of a correct inductive inference with true premises. Whether the inductive inference is correct depends upon whether the rule governing that inference is correct. The relation of inductive evidential support is, therefore, inseparably bound to the correctness of rules of inductive inference. In order to be able to say whether a given statement is supported by inductive evidence we must be able to say which inductive rules are correct.

For example, suppose that a die has been thrown a large number of times, and we have observed that the side two came up in one sixth of the tosses. This is our "evidence" e. Let h be the conclusion that, "in the long run," side two will come up one sixth of the times. Consider the following three rules:

1. (Induction by enumeration) Given m/n of observed A are B, to infer that the "long run" relative frequency of B among A is m/n.
2. (A priori rule) Regardless of observed frequencies, to infer that the "long run" relative frequency of B among A is $1/k$, where k is the number of possible outcomes—six in the case of the die.
3. (Counterinductive rule) Given m/n of observed A are B, to infer that the "long run" relative frequency of B among A is $(n - m)/n$.

Under Rule 1, e is positive evidence for h; under Rule 2, e is irrelevant to h; and under Rule 3, e is negative evidence for h. In order to say which conclusions are supported by what evidence, it is necessary to arrive at a decision as to what inductive rules are acceptable. If Rule 1 is correct, the evidence e supports the conclusion h. If Rule 2 is correct, we are justified in drawing the conclusion h, but this is entirely independent of the observational evidence e; the same conclusions would have been sanctioned by Rule 2 regardless of observational evidence. If Rule 3 is correct, we are not only prohibited from drawing the conclusion h, but also we are permitted to draw a conclusion h' which is logically incompatible with h. Whether a given conclusion is *supported by evidence*—whether it would be *rational to believe* it on the basis of given evidence—whether it is *made probable* by virtue of its relation to given evidence—depends upon selection of the correct rule or rules from among the infinitely many rules we might conceivably adopt.

The problem of induction can now be reformulated as a problem about evidence. What rules ought we to adopt to determine the nature of inductive evidence? What rules provide suitable concepts of inductive evidence? If we take the customary inductive rules to define the concept of inductive evidence, have we adopted a proper concept of evidence? Would the adoption of some alternative inductive rules provide a more suitable concept of evidence? These are genuine questions which need to be answered.[73]

We find, moreover, that what appeared earlier as a pointless question now becomes significant and difficult. If we take the customary rules of inductive inference to provide a suitable definition of the relation of inductive evidential support, it makes considerable sense to ask whether it is rational to believe on the basis of evidence as thus defined rather than to believe on the basis of evidence as defined according to other rules. For instance, I believe that the a priori rule and the counterinductive rule mentioned above are demonstrably unsatisfactory, and hence, they demonstrably fail to provide a suitable concept of inductive evidence. The important point is that something concerning the selection from among possible rules needs demonstration and is amenable to demonstration.

There is danger of being taken in by an easy equivocation. One meaning we may assign to the concept of inductive evidence is, roughly, the basis on which we ought to fashion our beliefs. Another meaning results from the relation of evidential support determined by whatever rule of inductive inference we adopt. It is only by supposing that these two concepts are the same that we suppose the problem of induction to have vanished. The problem of induction is still there; it is the problem of providing adequate grounds for the selection of inductive rules. We want the relation of evidential support determined by these rules to yield a concept of inductive evidence which is, in fact, the basis on which we ought to fashion our beliefs.[74]

We began this initially promising approach to the problem of the justification of induction by introducing the notion of probability, but we end with a dilemma. If we take "probability" in the frequency sense, it is easy to see why it is advisable to accept probable conclusions in preference to improbable ones. In so doing we shall be right more often. Unfortunately, we cannot show that inferences conducted according to any particular rule establish conclusions that are probable in this sense. If we take "probability" in a nonfrequency sense it may be easy to show that inferences which conform to our accepted inductive rules establish their conclusions as probable. Unfortunately, we can find no reason to prefer conclusions which are probable in this sense to those that are improbable. As Hume has shown, we have no reason to suppose that probable conclusions will often be true and improbable ones will seldom be true. This dilemma is Hume's problem of induction all over again. We have been led to an interesting reformulation, but it is only a reformulation and not a solution.

8. *Pragmatic Justification.* Of all the solutions and dissolutions proposed to deal with Hume's problem of induction, Hans Reichenbach's attempt to provide a pragmatic justification seems to me the most fruitful and promising.[75] This approach accepts Hume's arguments up to the point of agreeing that it is impossible to establish, either deductively or inductively, that any inductive inferences will ever again have true conclusions. Nevertheless, Reichenbach claims, the standard method of inductive generalization can be justified. Although its *success* as a method of prediction cannot be established in advance, it can be shown to be superior to any alternative method of prediction.

The argument can be put rather simply. Nature may be sufficiently uniform in suitable respects for us to make successful inductive inferences from the observed to the unobserved. On the other hand, for all we know, she may not. Hume has shown that we cannot prove in advance which case holds. All we can say is that nature may or may not be uniform—if she is, induction works; if she is not, induction fails. Even in the face of our ignorance about the uniformity of nature, we can ask what would happen if we adopted some radically different method of inference. Consider, for instance, the method of the crystal gazer. Since we do not know whether nature is uniform or not, we must consider both possibilities. If nature is uniform, the method of crystal gazing might work successfully, or it might fail. We cannot prove a priori that it will not work. At the same time, we cannot prove a priori that it will work, even if nature exhibits a high degree of uniformity. Thus, in case nature is reasonably uniform, the standard inductive method *must* work while the alternative method of crystal gazing *may or may not* work. In this case, the superiority of the standard inductive method is evident. Now, suppose nature lacks uniformity to such a degree that the standard inductive method is a complete failure. In this case, Reichenbach argues, the alternative method must likewise fail. Suppose it did not fail—

suppose, for instance, that the method of crystal gazing worked consistently. This would constitute an important relevant uniformity that could be exploited inductively. If a crystal gazer had consistently predicted future occurrences, we could infer inductively that he has a method of prediction that will enjoy continued success. The inductive method would, in this way, share the success of the method of crystal gazing, and would therefore be, contrary to hypothesis, successful. Hence, Reichenbach concludes, the standard inductive method will be successful *if any other method could succeed.* As a result, we have everything to gain and nothing to lose by adopting the inductive method. If any method works, induction works. If we adopt the inductive method and it fails, we have lost nothing, for any other method we might have adopted would likewise have failed. Reichenbach does not claim to prove that nature is uniform, or that the standard inductive method will be successful. He does not postulate the uniformity of nature. He tries to show that the inductive method is the best method for ampliative inference, whether it turns out to be successful or not.

This ingenious argument, although extremely suggestive, is ultimately unsatisfactory. As I have just presented it, it is impossibly vague. I have not specified the nature of the standard inductive method. I have not stated with any exactness what constitutes success for the inductive method or any other. Moreover, the uniformity of nature is not an all-or-none affair. Nature appears to be uniform to some extent and also to be lacking in uniformity to some degree. As we have already seen, it is not easy to state a principle of uniformity that is strong enough to assure the success of inductive inference and weak enough to be plausible. The vagueness of the foregoing argument is not, however, its fundamental drawback. It can be made precise, and I shall do so below in connection with the discussion of the frequency interpretation of probabililty.[76] When it is made precise, as we shall see, it suffers the serious defect of equally justifying too wide a variety of rules for ampliative inference.

I have presented Reichenbach's argument rather loosely in order to make intuitively clear its basic strategy. The sense in which it is a pragmatic justification should be clear. Unlike many authors who have sought a justification of induction, Reichenbach does not try to prove the truth of any synthetic proposition. He recognizes that the problem concerns the justification of a rule, and rules are neither true nor false. Hence, he tries to show that the adoption of a standard inductive rule is practically useful in the attempt to learn about and deal with the unobserved. He maintains that this can be shown even though we cannot prove the truth of the assertion that inductive methods will lead to predictive success. This pragmatic aspect is, it seems to me, the source of the fertility of Reichenbach's approach. Even though his argument does not constitute an adequate justification of induction, it seems to me to provide a valid core from which we may attempt to develop a more satisfactory justification.

III. Significance of the Problem

Hume's problem of induction evokes, understandably, a wide variety of reactions. It is not difficult to appreciate the response of the man engaged in active scientific research or practical affairs who says, in effect, "Don't bother me with these silly puzzles; I'm too busy doing science, building bridges, or managing affairs of state." No one, including Hume, seriously suggests any suspension of scientific investigation or practical decision pending a solution of the problem of induction. The problem concerns the *foundations* of science. As Hume eloquently remarks in *Enquiry Concerning Human Understanding:*

> Let the course of things be allowed hitherto ever so regular; that alone, without some new argument or inference, proves not that, for the future, it will continue so. In vain do you pretend to have learned the nature of bodies from your past experience. Their secret nature, and consequently all their effects and influence, may change, without any change in their sensible qualities. This happens sometimes, and with regard to some objects: Why may it not happen always, and with regard to all objects? What logic, what process of argument secures you against this supposition? My practice, you say, refutes my doubts. But you mistake the purport of my question. As an agent, I am quite satisfied in the point; but as a philosopher, who has some share of curiosity, I will not say scepticism, I want to learn the foundation of this inference.

We should know by now that the foundations of a subject are usually established long after the subject has been well developed, not before. To suppose otherwise would be a glaring example of "naïve first-things-firstism." [77]

Nevertheless, there is something intellectually disquieting about a serious gap in the foundations of a discipline, and it is especially disquieting when the discipline in question is so broad as to include the whole of empirical science, all of its applications, and indeed, all of common sense. As human beings we pride ourselves on rationality—so much so that for centuries rationality was enshrined as the very essence of humanity and the characteristic that distinguishes man from the lower brutes. Questionable as such pride may be, our intellectual consciences should be troubled by a gaping lacuna in the structure of our knowledge and the foundations of scientific inference. I do not mean to suggest that the structure of empirical science is teetering because of foundational difficulties; the architectural metaphor is really quite inappropriate. I do suggest that intellectural integrity requires that foundational problems not be ignored.

Each of two opposing attitudes has its own immediate appeal. One of these claims that the scientific method is so obviously the correct method that there is no need to waste our time trying to show that this is so. There are two difficulties. First, we have enough painful experience to know that the appeal to obviousness is dangerously likely to be an appeal to prejudice and superstition. What is obvious to one age or culture may well turn out, on closer examination, to be just plain false. Second, if the method of science is so obviously superior to other methods we might adopt, then I should think we ought to be able to point to those characteristics of the method by which it gains its obvious superiority.

The second tempting attitude is one of pessimism. In the face of Hume's arguments and the failure of many attempts to solve the problem, it is easy to conclude that the problem is hopeless. Whether motivated by Hume's arguments or, as is probably more often the case, by simple impatience with foundational problems, this attitude seems quite widespread. It is often expressed by the formula that science is, at bottom, a matter of faith. While it is no part of my purpose to launch a wholesale attack on faith as such, this attitude toward the foundations of scientific inference is unsatisfactory. The crucial fact is that science makes a *cognitive claim,* and this cognitive claim is a fundamental part of the rationale for doing science at all. Hume has presented us with a serious challenge to that cognitive claim. If we cannot legitimize the cognitive claim, it is difficult to see what reason remains for doing science. Why not turn to voodoo, which would be simpler, cheaper, less time consuming, and more fun?

If science is basically a matter of faith, then the scientific faith exists on a par with other faiths. Although we may be culturally conditioned to accept this faith, others are not. Science has no ground on which to maintain its *cognitive* superiority to any form of irrationalism, however repugnant. This situation is, it seems to me, intellectually and socially undesirable. We have had enough experience with various forms of irrationalism to recognize the importance of being able to distinguish them logically from genuine science. I find it intolerable to suppose that a theory of biological evolution, supported as it is by extensive scientific evidence, has no more rational foundation than has its rejection by ignorant fundamentalists. I, too, have faith that the scientific method is especially well suited for establishing knowledge of the unobserved, but I believe this faith should be justified. It seems to me extremely important that some people should earnestly seek a solution to this problem concerning the foundations of scientific inference.

One cannot say in advance what consequences will follow from a solution to a foundational problem. It would seem to depend largely upon the nature of the solution. But a discipline with well-laid foundations is surely far more satisfactory than one whose foundations are in doubt. We have only to compare the foundationally insecure calculus of the seventeenth and eighteenth centuries with the calculus of the late nineteenth century to appreciate the gains in elegance, simplicity, and rigor. Furthermore, the foundations of calculus provided a basis for a number of other developments, interesting in their own right and *greatly extending the power and fertility of the original theory.* Whether similar extensions will occur as a result of a satisfactory resolution of Hume's problem is a point on which it would be rash to hazard any prediction, but we know from experience that important consequences result from the most unexpected sources. The subsequent discussion of the foundations of probability will indicate directions in which some significant consequences may be found, but for the moment it will suffice to note that a serious concern for the solution of Hume's problem cannot fail to deepen our understanding of the nature of scientific inference. This, after all, is the ultimate goal of the whole enterprise.

IV. The Philosophical Problem of Probability

The foregoing lengthy discussion of the problem of induction has been presented, not only for its own sake, but also for its crucial bearing upon the problem of explicating the concept of probability. Although I cannot claim to have provided an exhaustive discussion of the whole variety of ways in which philosophers have tried to solve or dissolve Hume's problem, I do maintain that no such attempt has yet proved completely satisfactory. At the very least, there is nothing approaching universal agreement that any has succeeded. I have attempted to show, moreover, that the problem of induction does not immediately dissolve when the concept of probability is brought into the discussion. In the remaining sections I shall pursue the relation between probability and induction and try to show that the problem of induction not only does not vanish in the face of the probability concept, but rather poses the most fundamental problem that plagues our attempts to provide an intelligible explication of it. In order to elaborate this thesis, I shall present three general criteria of adequacy of probability concepts, explain why they are important, and examine certain leading theories in the light of them. To clarify the significance of these criteria, it is necessary to begin with a brief discussion of the mathematical calculus of probability.

As I explained earlier, the mathematical theory of probability had its serious beginnings in the seventeenth century through the work of Pascal and Fermat. Under the guidance of various empirical and intuitive considerations, the mathematical theory developed into a powerful tool with important applications in virtually every branch of science as well as in pursuits like poker, business, and war. In the present century a number of axiomatic constructions of the calculus of probability have been given.[78] These axiomatic constructions can be regarded as abstract formal systems of pure mathematics.[79] A formal system or abstract calculus consists of a set of formulas. Some of these formulas are singled out and taken as primitive; they are known as *axioms*. The remaining formulas of the system—the *theorems*—can be deduced purely formally from the axioms. The axioms are unproved and unprovable within the system; in fact, they are strictly meaningless, as are all the theorems derived from them. The axioms are meaningless because they contain *primitive terms* to which, from the standpoint of the formal system, no meaning is assigned. The formal system, as such, is concerned only with the deductive relations among the formulas; truth and falsity have no relevance to the formulas. Such systems with undefined primitive terms are said to be *uninterpreted*.

The probability calculus can be set up as a formal system in which the only primitive undefined term is that which stands for probability. All other terms in the calculus have well-established meanings from other branches of mathematics or logic. They provide, so to speak, the logical apparatus for the deduction of theorems from the axioms.

Psychologically speaking, the formal system is constructed with one eye on possible meanings for the primitive terms, but logically speaking, these considerations are no part of the formal system. Formal systems are, however, subject to interpretation. An *interpretation* consists in an assignment of meanings to the primitive terms. Two kinds of interpretation are possible, abstract and physical. An abstract interpretation is one that renders the system meaningful by reference to some other branch of mathematics or logic, and it makes the formulas into statements about the abstract entities in that domain. For example, Euclidean plane geometry can be axiomatized. When the primitive term "point" is interpreted so as to refer to pairs of numbers, and the primitive term "straight line" is made to correspond to certain classes of pairs of numbers, the result is analytic geometry. This exemplifies the notion of an abstract interpretation. A physical interpretation, in contrast, renders the primitive terms, and consequently the whole system, meaningful through reference to some part of the physical world. When, for example, straight lines are interpreted in terms of light rays, and points in terms of tiny pieces of metal, physical geometry is the result. Whether the interpretation is physical or abstract, the specification of meanings makes the formulas of the formal system into statements that are either true or false with regard to the entities of some domain of interpretation. Because of the deductive nature of the formal system, any interpretation that makes the axioms into true statements will make the theorems as well into true statements. It is through physical interpretation that formal systems achieve applicability to physical reality and utility for empirical science.

1. The Probability Calculus. For the sake of definiteness in discussing the various candidates for interpretations of the probability calculus, as well as for use in later discussions, I shall present a set of axioms for the elementary calculus of probability and derive some simple results.[80] The axioms are not minimal, for they contain redundancies; they have been chosen for intuitive clarity and ease of derivation of theorems rather than for mathematical elegance. I urge the reader with little mathematical background to forebear and be patient and to try to follow the discussion. It presupposes nothing more difficult than elementary arithmetic, and all of the derivations are extremely easy. I shall, moreover, provide concrete illustrations for the axioms and theorems.

Since probability is a relational concept, we shall incorporate its relational character into the axioms. Probability will, therefore, be a two-place function. Since the easiest examples are from games of chance, we can think of probability as relating types of events such as tosses of dice, spins of roulette wheels, drawing the queen of diamonds, or having three cherries show in a Las Vegas slot machine. The probability symbol, "$P(\, , \,)$" thus has two blanks within the parentheses, one before the comma and one after. For purposes of illustration we may think of symbols representing classes as the appropriate kind of symbol to insert in these places. We shall use the capital letters toward the beginning of the alphabet, "A," "B," "C," . . . to stand for classes. Thus, "$P(A, B)$" is a typical probability expression, and it represents the probability *from A to B*—the probability, given A, of getting B. If A is the class of tosses of a certain die, and B the class of cases in which side six lands uppermost, then $P(A, B)$ is the probability of getting six when you toss that die. The value of a probability is some number between zero and one, inclusive. In fact, the probability symbol simply stands for a number, and these numbers can be added, multiplied, etc. in the usual arithmetical way. Hence, the operations used to combine probabilities are the familiar arithmetical ones. Within the probability expression, however, we have symbols standing for classes; these can be manipulated by simple logical techniques. Symbols for a few class operations are required.[81] "$A \cup B$" stands for the *union* of A and B, that is, the class consisting of anything that belongs to A or B or both. "$A \cap B$" represents the *intersection* of A and B, that is, the class of things that belong to both A and B. "$\bar{A}$" designates the *complement* of A, the class containing everything not belonging to A. The capital Greek lambda "Λ" is the symbol for the *null class,* the class that has no members. We say that A is a *subclass* of B if every member of A is also a member of B. We say that two classes are *mutually exclusive* if they have no members in common—i.e., if their intersection is the null class.

The axioms may now be presented:

A1. $P(A,B)$ is a single-valued real function such that $0 \leq P(A,B) \leq 1$.
A2. If A is a subclass of B, $P(A, B) = 1$.
A3. If B and C are mutually exclusive $P(A, B \cup C) = P(A, B) + P(A,C)$.
A4. $P(A, B \cap C) = P(A, B) \times P(A \cap B, C)$

Axiom 1 tells us that a probability is a *unique* real number in the interval zero to one (including these endpoints). Axiom 2 states that the probability of an A being a B is one if every A is a B. The remaining two axioms are best explained by concrete illustration. Axiom 3 tells us, for example, that the probability of drawing a black card from a standard deck equals the probability of drawing a club plus the probability of drawing a spade. It does not apply, however, to the probability of drawing a spade or a face card, for these two classes are not mutually exclusive. Axiom 4 applies when, for example, we want to compute the probability of drawing two white balls in succession from an urn containing three white and three black balls. We assume for illustration that the balls are all equally likely to be drawn and that the second draw is made without replacing the first ball drawn. The probability of getting a white ball on the first draw is one half; the probability of getting a white ball on the second draw *if the first draw resulted in white* is two fifths, for with one white ball removed, there remain two whites and three blacks. The probability of two whites in succession is the product of the two probabilities—i.e., one fifth. If the first ball drawn is replaced before the second draw, the probability of drawing two whites is one fourth.

I shall now derive four rather immediate theorems.

T1. $P(A, \bar{B}) = 1 - P(A, B)$.

Proof: By axiom 2,

$$P(A, B \cup \bar{B}) = 1,$$

for every member of A is either a member of B or not a member of B. Hence, A is a subclass of $B \cup \bar{B}$. Since nothing is both a member of B and not a member of B, B and $\bar{B}$ are mutually exclusive. By axiom 3,

$$P(A, B \cup \bar{B}) = P(A, B) + P(A, \bar{B}) = 1.$$

Theorem 1 follows by subtraction.

T2. $P(A, \Lambda) = 0$.

Proof: Since nothing is both B and $\bar{B}$, $B \cap \bar{B} = \Lambda$, the null class. But the complement, $\overline{B \cap \bar{B}}$, of the null class is the class $B \cup \bar{B}$ that contains everything that either is a B or is not a B. Substituting in theorem 1, we have

$$P(A, B \cap \bar{B}) = 1 - P(A, \overline{B \cap \bar{B}}) = 1 - P(A, B \cup \bar{B}).$$

Using the fact, established in the proof of theorem 1, that $P(A, B \cup \bar{B}) = 1$, theorem 2 follows.

T3. $P(A, C) = P(A, B) \times P(A \cap B, C) + P(A, \bar{B}) \times P(A \cap \bar{B}, C)$.

Proof: The class of things that belong to the class C is obviously the class of things that are both B and C or else both $\bar{B}$ and C; hence,

$$P(A, C) = P(A, [B \cap C] \cup [\bar{B} \cap C]).$$

Moreover, since nothing can be a member of both B and $\bar{B}$, $B \cap C$ and $\bar{B} \cap C$ are mutually exclusive. Axiom 3 therefore yields

$$P(A, [B \cap C] \cup [\bar{B} \cap C]) = P(A, B \cap C) + P(A, \bar{B} \cap C).$$

Axiom 4 gives

$$P(A, B \cap C) = P(A, B) \times P(A \cap B, C)$$

and

$$P(A, \bar{B} \cap C) = P(A, \bar{B}) \times P(A \cap \bar{B}, C).$$

Combining these results we get theorem 3.

T4. If $P(A, C) \neq 0$,

$$P(A \cap C, B) = \frac{P(A, B) \times P(A \cap B, C)}{P(A, C)}$$

$$= \frac{P(A, B) \times P(A \cap B, C)}{P(A, B) \times P(A \cap B, C) + P(A, \bar{B}) \times P(A \cap \bar{B}, C)}.$$

Proof: By axiom 4,

$$P(A, C \cap B) = P(A, C) \times P(A \cap C, B).$$

Therefore,

$$P(A \cap C, B) = \frac{P(A, C \cap B)}{P(A, C)} \tag{1}$$

if $P(A, C) \neq 0$. Since $B \cap C$ is obviously the same class as $C \cap B$, we have, using axiom 4,

$$P(A, C \cap B) = P(A, B \cap C) = P(A, B) \times P(A \cap B, C).$$

Using this result to replace the numerator in (1) establishes the first equality. Since theorem 3 equates the denominator of the middle member of the theorem with that of the third member, the entire theorem is established. Theorem 4 is a simple form of Bayes' theorem, and it will be used extensively in the discussion of subsequent problems.

I shall now present a concrete example of the application of each of the four theorems.

Theorem 1. The probability of throwing some number other than six with a standard die is equal to one minus the probability of getting six. This theorem, although extremely simple and obvious, is rather useful. Consider the probability of tossing at least one six in three trials—i.e., the probability of getting six on the first or second or third toss. Since these events are not mutually exclusive, axiom 3 does not apply. It is possible to calculate this probability directly, but it is much easier to do so by calculating the probability of its nonoccurrence and then using theorem 1. By two applications of axiom 4, we find that the probability of getting nonsix on the first *and* second *and* third tosses is $5/6 \times 5/6 \times 5/6 = 125/216$. Subtracting this value from one, we find that the probability of getting at least one six in three tosses is $91/216$ (which is considerably smaller than $1/2$). One of the problems posed by Chevalier de Méré that led to the beginning of the mathematical theory of probability is a more complicated instance of the same problem. The famous gentleman asked, "How many times must one toss a pair of fair dice in order to have at least a fifty-fifty chance of getting at least one double-six?" The naïvely tempting answer "18" is obviously wrong; it is by no means immediately evident what the right answer is.[82]

Theorem 2. This is obvious: The probability of getting a result that is both odd and even on a toss of a die is zero.

Theorem 3. Consider the following game of chance. Let there be two decks of cards, each containing twelve cards. The first deck contains two red cards and ten black; the second contains six red and six black. The player must first roll a die to determine from which deck to draw. If the side one shows, he draws from the deck containing six red cards; if any other side shows, he must draw from the deck with only two red cards. The draw of a red card constitutes a win. We can use theorem 3, which is known as the *theorem on total probability,* to compute the player's chance of winning. Let A be rolls of the die (which is tantamount to playing), let B be the side one showing (which is tantamount to drawing from the deck containing six red cards), and let C be drawing a red card. Substituting the appropriate values in the theorem yields

$$\frac{2}{9} = \frac{1}{6} \times \frac{1}{2} + \frac{5}{6} \times \frac{1}{6}.$$

The player's chance of winning is two ninths.

Theorem 4. Continuing with the game used to illustrate theorem 3, suppose a player has just made a winning draw, but we did not notice from which deck it came. What is the probability that he drew it from the half-red deck? Bayes' theorem supplies the answer. Using the result just established by theorem 3, we find that the probability that the win came from the half-red deck is

$$\frac{3}{8} = \frac{\frac{1}{6} \times \frac{1}{2}}{\frac{2}{9}}.$$

Notice that the probability of the winning card coming from that deck is less than one half, although the probability of getting a red card if you draw from that deck is considerably greater than the probability of getting a red card if you draw from the other deck. Given that a winning draw has occurred, the chances favor its having come from the predominantly black deck simply because the vast majority of draws are made from that deck.

Although a number of concrete examples have been presented to provide intuitive understanding of the axioms and theorems, these illustrations constitute no part of the formal calculus of probability itself. The proofs of theorems depend in no way upon the examples; they are carried out strictly formally and are entirely independent of any interpretation.

The foregoing discussion of the elementary calculus of probability provides a sufficient basis to proceed with the problem of interpreting the calculus. This is, I take it, *the* fundamental *philosophical problem of probability.* It is the problem of finding one or more interpretations of the probability calculus that yield a concept of probability, or several concepts of probability, which do justice to the important applications of probability in empirical science and in practical affairs. Such interpretations, whether one or several, would provide an explication of the familiar notion of probability. It is perhaps worth mentioning at this point that the problem is not one of empirical or quasi-empirical linguistics. We are not primarily concerned with the ways a language user, whether the man in the street or the scientist, uses the English word "probability" and its cognates. We are concerned with the logical structure of science, and we need to provide concepts to fulfill various functions within that structure. If the word "probability" and its synonyms had never occurred in any language, we would still need the concept for the purpose of logical analysis. As Carnap has remarked, if it did not already exist we would have to invent it. Moreover, the difficulties we shall discover—centering mainly on Hume's problem of induction—are ones that arise out of subtle philosophical analysis. Ordinary people are not aware of these problems, so ordinary usage cannot be expected to be sensitive to them.

2. *Criteria of Adequacy for Interpretations.* In order to facilitate our investigation of the philosophical problem of probability, I shall state three criteria which must be fulfilled, I believe, if we are to have a satisfactory interpretation of probability. Although the criteria seem simple and straightforward, we shall see that it is exceedingly difficult to find any interpretation that satisfies all three.

a) *Admissibility.* We say that an interpretation of a formal system is admissible if the meanings assigned to the primitive terms in this interpretation transform the formal axioms, and consequently all the theorems, into true statements. A fundamental requirement for probability concepts is to satisfy the mathematical relations specified by the calculus of probability. This criterion is not merely an expression of admiration of mathematics; important reasons can be given for insisting upon it. One reason is that the mathematical calculus has been developed with great care and precision over a long period of time and with due regard for a vast range of practical and theoretical problems. It would be rash indeed to conclude on the basis of casual reflection that the mathematical theory is likely to be wrong or irrelevant in relation to potential applications. Another reason for insisting upon admissibility is a consequence of the fact that violations of the formal properties of the calculus lead to *incoherent betting systems.* This consideration figures crucially in the personalistic interpretation of probability, which will be discussed below. For the moment, an example will suffice. Suppose, for instance, that someone were to violate theorem 1 and axiom 2 by assigning probability values that do not add up to one for all possible outcomes. He assigns, say, the value one third to the probability of heads and one third to the probability of tails on tosses of a coin whose outcome must be one or the other of these. If it lands on edge, we do not count that toss. Such a person would presumably give odds of two to one that the coin will not come up heads, and he will also give odds of two to one that it will not come up tails. If he makes both of these bets, however, he is bound to lose whatever happens. If it comes up heads he wins one dollar and loses two, and if it comes up tails the result is the same. Anyone who wants to use probabilities to determine betting odds must guard against such situations. It has been shown that satisfaction of the admissibility criterion is a necessary and sufficient condition for the avoidance of incoherent betting systems.[83]

b) *Ascertainability.* This criterion requires that there be some method by which, in principle at least, we can ascertain values of probabilities. It merely expresses the fact that a concept of probability will be useless if it is impossible in principle to find out what the probabilities are.[84]

c) *Applicability.* The force of this criterion is best expressed in Bishop Butler's famous aphorism, "Probability is the very guide of life." [85] It is an unescapable fact that we are seeking a concept of probability that will have practical predictive significance. For instance, knowledge of the probabilities associated with throws of dice should have an important bearing upon the kinds of bets we are willing to make. Knowledge of the probability of radioactive decay should have some bearing upon our prediction of the amount of a given substance that will remain undecayed after a certain time.

More generally, it appears that one or more probability concepts play fundamental roles in the logical structure of science. There are, for instance, statistical or probabilistic laws in science. The second law of thermodynamics, that in a closed system with a low entropy state the entropy will very probably increase, is a leading example. Any science that attempts precise measurement must deal with errors; the concept of error is basically probabilistic. Moreover, the results of scientific inference, in some important sense, are probable. Since scientific inference is ampliative, its conclusions do not follow with necessity or certainty from the data. The concept of scientific confirmation is another example of a fundamental scientific concept that is unavoidably probabilistic.

The foregoing remarks indicate, at least roughly, some of the functions our explication of probability must fulfill and some of the contexts into which it (or they) must fit. The force of the criterion of applicability is merely to call attention to these functions. An explication that fails to fulfill the criterion of applicability is simply not an explication of the concept we are trying to explicate.

It may now be obvious that the fundamental philosophical difficulty in the theory of probability lies in the attempt to satisfy simultaneously the criteria of ascertainability and applicability. Perhaps it is also obvious that this is Hume's problem of induction all over again in slightly different terminology. If these points are not obvious, I hope they will become so as we consider possible candidates for the interpretation of the probability concept.

V. Interpretations of Probability

This section will survey five leading interpretations of probability, confronting each of them with the three foregoing criteria.

1. The Classical Interpretation. This interpretation is one of the oldest and best known; it defines probability as the ratio of favorable to equally possible cases.[86] With a perfectly symmetrical die, for instance, the probability of tossing an even number is three sixths. Three sides have even numbers—the favorable cases—and there are six equally possible sides. The immediate difficulty with this interpretation is that "equally possible" seems to mean "equally probable," so the definition appears to be flagrantly circular. But the apparent circularity can be overcome if a definition of "equally probable" can be given which is independent of the definition of "probable" itself. The classical theorists attempted to offer such a definition by means of the *principle of indifference.* This principle states that two possibilities are equally probable if there is no reason to prefer one to the other.

The principle of indifference, lying as it does at the very heart of the classical interpretation, has been the subject of much controversy. Various objections have been brought against it.[87] First, it *defines* "probability" in terms of equally probable alternatives, so it presupposes a priori that every instance of probability can be analyzed in terms of equally probable cases. Suppose, for example, that we have a slightly biased coin—one for which the probability of heads is 0.51 and the probability of tails is 0.49. How are we to find the 100 equally probable occurrences, of which 51 are favorable to heads? Analogously, since the birth rate for boys slightly exceeds the birth rate for girls, how can we be assured a priori that we can find an appropriate number of equally probable alternatives in the genetic mechanism, of which a suitable number result in the birth of a boy? To suppose it is always possible to reduce unequal probabilities to sets of equiprobable cases is a rash and unwarranted assumption.

Another objection rejects any rule that pretends to transform ignorance automatically into knowledge. Knowledge of probabilities is concrete knowledge about occurrences; otherwise, it is useless for prediction and action. According to the principle of indifference, this kind of knowledge can result immediately from our ignorance of reasons to regard one occurrence as more probable than another. This is epistemological magic. Of course, there are ways of transforming ignorance into knowledge—by further investigation and the accumulation of more information. It is the same with all "magic"; to get the rabbit out of the hat you first have to put him in. The principle of indifference tries to perform "*real* magic."

The decisive objection against the principle shows that it gives rise to explicit logical contradiction.[88] Consider a simple example. Suppose we know that a car has taken between one and two minutes to traverse one mile, and we know nothing further about the amount of time taken. Applying the principle of indifference, we conclude that the probability that the time was between one and one-and-one-half minutes equals the probability that it was between one-and-one-half and two minutes. Our data can, however, be expressed in another way. We know that the *average* speed for the trip was between sixty and thirty miles per hour, but we know nothing further about the average speed. Applying the principle of indifference again, we conclude that the probability that the average speed was between sixty and forty-five miles per hour equals the probability that it was between forty-five and thirty miles per hour. Unfortunately, we have just contradicted the first result, because the time

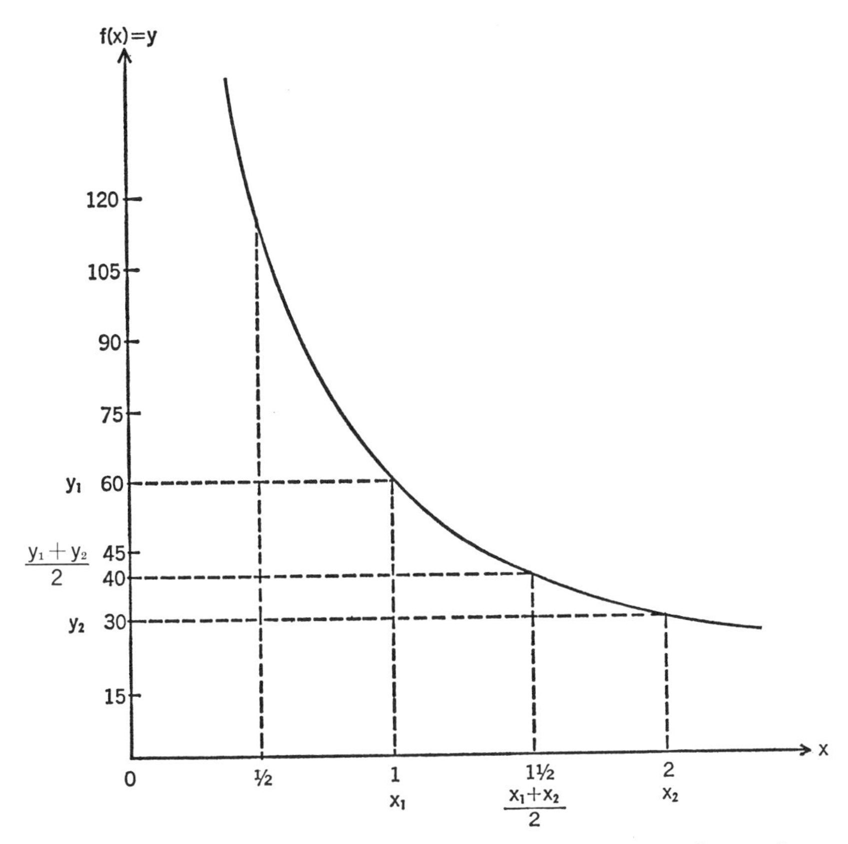

Figure 1. Graph of time and average speed for the fixed distance of one mile.

of one-and-one-half minutes corresponds to forty, not forty-five, miles per hour.

This example is not just an isolated case (although one contradiction ought to be enough), but it illustrates a general difficulty. A similar conflict can be manufactured out of any situation in which we have two magnitudes interdefined in such a way that there is a nonlinear relation between them. For a fixed distance, speed and time are reciprocal; for a square, area relates to the second power of the length of the side; etc. Figure 1 shows graphically the source of the paradox. For a nonlinear function, the value of the function for the *mid-point* $(x_1 + x_2)/2$ *between two values of the argument* x does not, in general, correspond to the *mid-point* $(y_1 + y_2)/2$ *between the values of the function* determined by those two values of the argument.

The whole set of paradoxes generated in the foregoing manner shows that the principle of indifference yields equivocal values for probabilities. The probability of averaging at least forty-five miles per hour over the mile is equal to one half and also to some number other than one half. Since axiom 1 of the probability calculus requires that probability values be unique, the classical interpretation violates the criterion of admissibility.

2. *The Subjective Interpretation.* Some authors have spoken as if probability is simply a measure of degree of belief.[89] If I believe with complete certitude that the sun will rise tomorrow, my degree of belief is one and so is my subjective probability. If I am just as strongly convinced that a penny will turn up heads as that it will not turn up heads, my degree of belief and subjective probability for each of these outcomes is one half. Leaving aside, for now, the very serious question of the adequacy of a purely subjective concept for application within empirical science, we can easily see that this interpretation fails to satisfy the criterion of admissibility. Certain seventeenth-century gamblers, for example, believed to a degree one thirty-sixth in each of the possible outcomes of a throw of two dice, and this degree was unaffected by the outcomes of the preceding throws. Moreover, these same men believed more than they disbelieved in getting at least one double six in twenty-four throws. It was the Chevalier de Méré's doubts that led him to approach Pascal with his famous problem. Computation showed, as it turned out, that the probability of getting double six, under the conditions specified, is less than one half. Since the degrees of belief do not coincide with the calculated values, the subjective interpretation does not constitute an admissible interpretation of the probability calculus.

3. *The Logical Interpretation.* Although we cannot interpret probability as degree of *actual* belief, it might still be possible to maintain that probability measures degree of *rational* belief. As a matter of fact, there is good reason to believe that many of the earlier authors who spoke as if they were adopting a purely subjective interpretation were actually much closer to an interpretation in terms of rational belief.[90] According to this interpretation, probability measures the degree of confidence that would be rationally justified by the available evidence. Although often formulated in terms of the psychological concept of belief, there is nothing at all subjective about this interpretation. Probability is regarded as an objective logical relation between statements that formulate evidence and other statements—*hypotheses*—whose truth or falsity is not fully determined by the evidence.[91]

According to the logical theory, there is a fundamental analogy

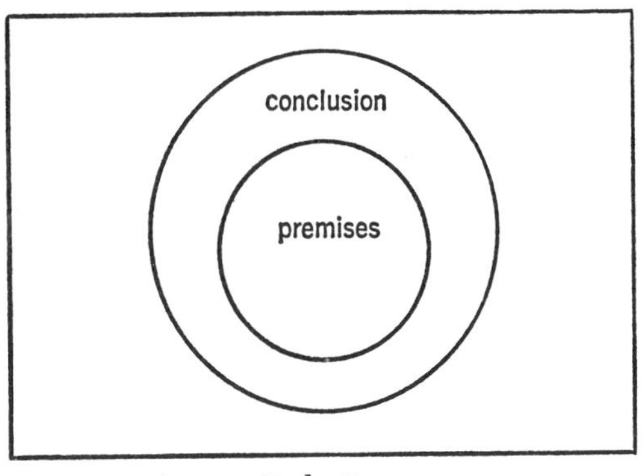

Deduction

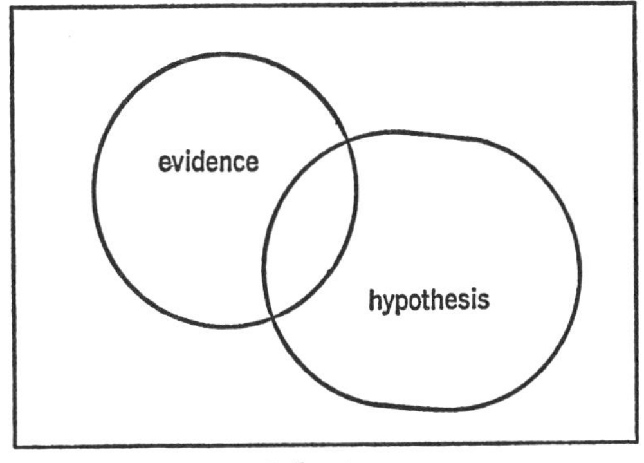

Induction

Figure 2

between inductive and deductive logic. Deductive logic embodies a concept of logical entailment between premises and conclusion. A conclusion that is logically entailed by true premises cannot be false. Inductive logic requires a logical concept of probability, also known as *degree of confirmation,* relating evidence to hypotheses. In this case, the evidence does not logically entail the hypothesis, for the hypothesis could be false even when the statements of evidence are true. But there is a relation of partial entailment, and this is what probability measures. The situation is shown diagrammatically in Figure 2. In each case, the rectangle represents all possible states of affairs. When premises entail a conclusion, every possible case in which the premises are true is a case in which the conclusion is also true; thus, the premise-circle is entirely contained within the conclusion-circle. When evidence partially entails a hypothesis, there is some degree of overlapping between those possible states of affairs in which the evidence holds and those in which the hypothesis is true. Degree of confirmation measures the amount of overlapping.

These considerations may be heuristically helpful, but the concept of logical probability must be made much more precise. In particular, we must specify what constitutes a possible state of affairs and how such things are to be counted or measured. We have a good understanding of how to measure areas; we must not be misled by geometrical diagrams into thinking we know how to measure degree of "overlapping" of possibilities or degree of partial entailment. This matter poses severe difficulties.

Rudolf Carnap, who has given the fullest account of degree of confirmation, proceeds by constructing a precise language and selecting the strongest descriptive statements that can be made in that language. Such languages are capable of describing various universes. A simple language can describe a simple universe; a more complex language is able to describe a more complex universe. Each of the strongest descriptive statements in a language gives a complete description of some possible state of the corresponding universe. Consider an extremely simple language. The illustrative language will have three terms, "*a*," "*b*," and "*c*," as names for individual entities, and one descriptive predicate "*F*" referring to a property these individuals may have or lack. In addition, it will have the usual logical apparatus. To have a concrete example, we may think of a universe that consists entirely of three balls, *a*, *b*, and *c*, such that the only property that can be meaningfully predicated of them is the property of being red. Although this universe, and the language used to describe it, are absurdly simple, they are sufficient to elicit many of the important properties of logical probability.

There are eight possible states of the universes describable by our language, depending upon which individuals have or lack the property red. Statements describing the possible states are *state descriptions*. Letting "Fa" mean "*a* is red," etc., "$\sim Fa$" mean "*a* is not red," etc., and the dot mean "and," the complete list of state descriptions is as follows:

1. $Fa \cdot Fb \cdot Fc$
2. $\sim Fa \cdot Fb \cdot Fc$
3. $Fa \cdot \sim Fb \cdot Fc$
4. $Fa \cdot Fb \cdot \sim Fc$
5. $\sim Fa \cdot \sim Fb \cdot Fc$
6. $\sim Fa \cdot Fb \cdot \sim Fc$
7. $Fa \cdot \sim Fb \cdot \sim Fc$
8. $\sim Fa \cdot \sim Fb \cdot \sim Fc$

Each state description is a conjunction of simple statements, each of which asserts or denies that a given individual has the property in question; moreover, each state description contains one such assertion or denial for each individual. Any statement that can be meaningfully formulated in this language can be expressed in terms of state descriptions. For instance, the statement, "There is at least one thing that has the property *F*," can be translated into the disjunction of state descriptions 1–7. The statement, "Individual *a* has property *F*," can be translated into the disjunction of state descriptions 1, 3, 4, and 7. The statement, "Everything has the property *F*," is equivalent to state description 1 by itself. In each of the foregoing examples, we say that the statement *holds in* the state descriptions mentioned, and the set of state descriptions in which a statement holds is called the *range* of the statement. The range of a statement consists of all the state descriptions that are logically compatible with it; those outside the range are incompatible with the statement in question.

Any statement that can be formulated in the language can serve as a hypothesis, and any consistent statement can serve as evidence. To determine the degree to which a hypothesis is supported by evidence, we turn to the ranges of these statements. In particular, we compare the range of the evidence statement with the portion of its range that coincides with the range of the hypothesis. For instance, we might take as our hypothesis the statement that everything has the property *F*, and as our evidence that individual *a* has that property. The evidence holds in four state descriptions, as noted above, while evidence and hypothesis together hold in only one state description, namely, the first.

At this point it is very tempting merely to count up the state descriptions in the two ranges and form their ratio. Letting "*e*" stand for evidence, "*h*" for hypothesis, and "*R*( )" for the number of state descriptions in the range of, we might hurriedly conclude that the degree of confirmation of *h* on *e* is given by

$$\frac{R(e \cdot h)}{R(e)} = \frac{1}{4}$$

This amounts to an assignment of equal weights to all of the state descriptions. Since it is convenient to have the weights of the state descriptions add up to one, equal weighting will mean that each state description has the weight one eighth.

The assignment of equal weights to the state descriptions may be regarded as an application of the principle of indifference to the fundamental possibilities in our model universe. It evades the paradox associated with the classical unrestricted application. The classical theory of probability got into trouble, one might say, because of a failure to distinguish fundamental possibilities from derivative possibilities, and because of the consequent failure to stipulate exactly what properties the principle applies to and what ones it does not. Restriction of the principle of indifference to the state descriptions provides precisely the specification needed to avoid contradiction. The explication of probability in these terms has been thought to preserve the "valid core" of the traditional principle of indifference.

Carnap has shown, however, that the assignment of equal weights, although intuitively plausible, is radically unsatisfactory. Such an assignment would have the consequence that it is logically impossible to learn from experience! Consider as an hypothesis h the statement "Fc." Leaving factual evidence completely aside, h holds in four state descriptions and not in the other four. Its probability, a priori, is one half. Suppose now that we begin to accumulate factual evidence; for example, suppose we examine a and find it has property F. The evidence alone holds in state descriptions 1, 3, 4, and 7, while evidence and hypothesis together hold in 1 and 3. The probability of our hypothesis, even after we have found some evidence, is still one half. This hypothesis retains the same probability regardless of what observations we make of individuals a and b, and this probability is absolutely insensitive to whether the evidence is what we would usually regard as favorable or unfavorable to the hypothesis. It is evident that we must find a more satisfactory way of weighting the state descriptions, and this will, of course, involve assignments of unequal weights to different state descriptions.

To circumvent the difficulty about learning from experience, Carnap proposed a different system for assigning weights to state descriptions which he designated "m^*." [92] He argued for its plausibility along the following lines. The only thing that significantly distinguishes individuals from one another is some qualitative difference, not a mere difference of name or identity. Hence, the fundamental possibilities to which we assign equal weights should not discriminate on the basis of individuals. Consider, for instance, state descriptions 2, 3, and 4. They are all alike in affirming that two of the individuals possess the property F while denying it to the third. These state descriptions differ from one another only in identifying the individual that lacks property F. Similar remarks apply to state descriptions 5, 6, and 7; all of them deny the property F to two individuals while asserting that it holds of one. Carnap says that state descriptions that differ from each other only in the arrangement of the individual names are *isomorphic;* a collection of all the state descriptions isomorphic to each other is a *structure description.* In our simple language there are four structure descriptions: state description 1 by itself; state descriptions 2, 3, and 4; state descriptions 5, 6, and 7; and state description 8 by itself. A structure description can be construed as the disjunction of all the state descriptions that make it up; for example, the second structure description asserts that state description 2, 3, *or* 4 obtains. Thus, in effect, a structure description tells us *how many* individuals have the properties in question, but not *which ones*. A state description, of course, tells us both.

Carnap's system m^* takes the structure descriptions as fundamental and assigns equal weights to them. In our example, the weight of each would be one fourth. The principle of indifference is applied to structure descriptions in the first instance. Then it is applied once again to the state descriptions within any given structure description. In our example, this establishes the weight one twelfth for state descriptions 2, 3, and 4. They must have equal weights which add up to one fourth, the total weight for the structure description they compose. State descriptions 5, 6, and 7 are given the weight one twelfth for the same reason. State descriptions 1 and 8 each receive the weight one fourth, for each constitutes a structure description by itself. The system can be summarized as follows:

State Description	*Weight*	*Structure Description*	*Weight*
1. $Fa \cdot Fb \cdot Fc$	one-fourth	I. All F	one-fourth
2. $\sim Fa \cdot Fb \cdot Fc$	one-twelfth		
3. $Fa \cdot \sim Fb \cdot Fc$	one-twelfth	II. $2\ F, 1 \sim F$	one-fourth
4. $Fa \cdot Fb \cdot \sim Fc$	one-twelfth		
5. $\sim Fa \cdot \sim Fb \cdot Fc$	one-twelfth		
6. $\sim Fa \cdot Fb \cdot \sim Fc$	one-twelfth	III. $1\ F, 2 \sim F$	one-fourth
7. $Fa \cdot \sim Fb \cdot \sim Fc$	one-twelfth		
8. $\sim Fa \cdot \sim Fb \cdot \sim Fc$	one-fourth	IV. No F	one-fourth

Let us now consider, as before, the hypothesis "Fc" in the light of various evidence statements, this time using a concept of confirmation based upon m^*. In each case we simply calculate $m^*(e \cdot h)/m^*(e)$. First we note that the a priori probability of our hypothesis on the basis of no factual evidence at all is one half, for it holds in state descriptions 1, 2, 3, and 5, whose combined weight is one half. Now, suppose we know that a has the property F; let this be evidence e_1. "$Fa \cdot Fc$" holds in state descriptions 1 and 3, whose combined weights, one fourth and one twelfth, equal one third. "Fa" alone holds in state descriptions 1, 3, 4, and 7, whose weights total one half. The degree of confirmation of our hypothesis on this evidence is

$$c^*(h, e_1) = \frac{m^*(e_1 \cdot h)}{m^*(e_1)} = \frac{1/3}{1/2} = \frac{2}{3}$$

Happily, the observation that individual a has property F, which would be regarded intuitively as positive evidence for the hypothesis that c has the property F, does raise the probability of that hypothesis. You may verify by your own computation that

for e_2 $(Fa \cdot Fb)$, $c^*(h, e_2)$ = three-fourths
for e_3 $(\sim Fa)$, $c^*(h, e_3)$ = one-third
for e_4 $(\sim Fa \cdot \sim Fb)$, $c^*(h, e_4)$ = one-fourth
for e_5 $(Fa \cdot \sim Fb)$, $c^*(h, e_5)$ = one-half

Clearly, m^* yields a concept of degree of confirmation c^* that allows for the possibility of learning from experience.

Many objections have been raised against the explication of degree of confirmation as c^*, most of which were well known to Carnap and discussed explicitly by him. In the years since the publication of his major treatments of c^*, he has himself sought improved systems of inductive logic and has abandoned c^* except as a useful approximation in certain contexts. For example, one of the most serious shortcomings of the system based on m^* (and of many related systems as well) is the fact that the degree of confirmation of a hypothesis depends in an unfortunate way upon the language in which the confirmation function is defined. The addition of another predicate "G"—standing for "hard," let us say—would alter the probabilities of the statement that c has the property F in relation to the evidence statements given above—in relation to evidence that does not involve the new property at all! [93] To circumvent this difficulty, Carnap originally imposed the requirement of descriptive completeness of the language, but the requirement is not plausible, and Carnap never defended it enthusiastically. His more recent work dispenses with this completeness requirement without reintroducing the same difficulty.

Another shortcoming of the system based on m^* (as well as many other similar systems) is that in a language with infinitely many names for individuals, no universal generalization of the form "All F are G" can ever have a degree of confirmation other than zero on the kind of evidence that can be accumulated by observation. Carnap has recently been working with systems that do not have this drawback.

Carnap's earlier systems of inductive logic have been criticized—especially by those who are more interested in questions of practical statistical inference than in foundational questions—on the ground that the confirmation functions were defined only for extremely simple languages. These languages embody only qualitative predicates; since they do not contain quantitative concepts, they are patently inadequate as languages for science. In his more recent work, Carnap has been developing systems that are able to treat physical magnitudes quantitatively.

As the preceding remarks should indicate, Carnap has been acutely aware of various technical difficulties in his earlier treatments of the logical interpretation of probability, and he has made enormous strides in overcoming them.[94] There is, however, in my opinion, a fundamental problem at the heart of the logical interpretation. It is, I think, a difficulty of principle which cannot be avoided by technical developments. It seems to be intrinsic to the entire conception of probability as a logical relation between evidence and hypothesis.

The logical interpretation involves, in an essential way, the conception of probability as a measure of possible states of affairs. Whether the measure is attached to statements describing these possibilities (i.e., state descriptions or structure descriptions) or to the possibilities themselves (i.e., facts, propositions, models), the measure is indispensable. There are many alternative ways of assigning such a measure; for instance, there are infinitely many different ways of assigning nonnegative weights to the state descriptions of our simple illustrative language in such a way that together they total one. As a matter of fact, Carnap has described a continuum of weightings, and there are others beyond the scope of that collection.[95] Alternative methods of weighting have, of course, differing degrees of intuitive plausibility. The inescapable question is: How are we to select the appropriate weighting or measure from the superdenumerable infinity of candidates?

One feature of the choice has signal importance: *The choice must be a priori.* We may not wait to see how frequently a property occurs in order to assign it a weight; an a posteriori method of this sort would have to find its place within one of the interpretations still to be discussed. The problem is to show how we can make a choice that is not completely arbitrary. Assuming we have made a definite choice, this choice *defines* the notions of degree of confirmation and logical probability. It determines our inductive logic. As a consequence, degree of confirmation statements, or statements of logical probability, are *analytic* if they are true (self-contradictory if they are false). They are statements whose truth or falsity are results of definitions and pure logic alone; they have no synthetic or factual content. Given any hypothesis h and any consistent evidence statement e, the degree of confirmation $c(h, e)$ can be established by computation alone, as we saw in dealing with our examples above. The question is: *How can statements that say nothing about any matters of fact serve as "a guide of life?"* [96]

Carnap's immediate answer is that we do not use these degree-of-confirmation statements by themselves. In an attempt to answer this very question, he points out that the evidence statement e is synthetic, and it is the analytic degree of confirmation statement *in combinaiion with the synthetic evidence statement* that provides a guide to decisions and action. Very well, analytic statements alone do not serve as "a guide of life." A problem still remains. In the cases that interest us, we are trying to deal with unobserved matters of fact on the basis of evidence concerning observed matters of fact. We are trying to make a prediction or decision regarding a future occurrence. The evidence statement, although synthetic, is not a statement about any future event; it has factual content, but that content refers entirely to the past. How, we may ask, can a synthetic statement *about the past,* in conjunction with an analytic degree of confirmation statement, tell us anything about the future? How, in particular, can it serve as any kind of guide to prediction, decision, and action?

We must not allow a fundamental misconception to cloud this issue. The way in which the evidence statement and the degree of confirmation statement function is very different from the role of premises in an inference. Given that the degree of confirmation of hypothesis h on evidence e is p, and given also the truth of e, we are not allowed to infer h even if p is very near one. Rather, we must use our inductive logic according to certain definite rules of application.[97] First, there is the *requirement of total evidence.* If e is the evidence statement we are going to use, it must incorporate all relevant available evidence. This is an important respect in which inductive logic differs from deductive. If a conclusion B follows deductively from a premise (or set of premises) A, and if we know that the premise is true, we may assert the conclusion B, even if we know a great deal that is not stated in A. If B follows validly from A, it also follows validly from A conjoined with any additional premises whatever. By contrast, a hypothesis h may be highly confirmed with respect to evidence e, but its degree of confirmation on the basis of e and additional evidence e' may be very low indeed.

Given, then, that e is all the available relevant evidence, we are still not allowed to assert h. Instead, we may use our confirmation statement, "$c(h, e) = p$" to determine what would constitute a reasonable bet on h.

For example, suppose (since the arithmetic is already done) that we have a collection of three objects, and all we know is that two of them have been observed, and they are both red. What sort of bet should we be willing to make on the third one being red? Leaving certain niceties aside, if we accept c^* as an adequate explication of degree of confirmation, we should be willing to give odds of three to one that the next one is red, for the degree of confirmation is three quarters.[98]

Let us look at this situation in a slightly more dramatic way. Suppose I am thinking about making a wager on the outcome of the next race; in particular, I am thinking of betting on Kentucky Kid. Not being especially knowledgeable regarding horse racing or pari-mutuel betting, I bring along an advisor who is well versed in logical probability. I ask him whether KK will win. He tells me he does not know and neither does anyone else, since the race is not fixed. I tell him I had not expected a categorical and certain reply, but that I would like some information about it in order to decide how to bet. Going over the racing form together, we accumulate all of the relevant information about KK's past performance, as well as that of the other horses in the race. I tell him I find it all very interesting, but what about the next race? After a rapid calculation he tells me that the probability of KK's winning, with respect to all of the available relevant evidence, is one half. Under these circumstances, he tells me, a bet at even odds is reasonable. I am about to thank him for his information about the next race, when I reflect that the degree of confirmation statement is analytic, while the evidence statement refers only to past races and not to the future one I am concerned with, so I cannot see that he has supplied me with any information about the future race. I voice my discontent, saying I had hoped he would tell me *something* about the next race, but he has only given me information about past races. He replies that he *has* told me what constitutes a fair bet or a *rational degree of belief*. Surely, he says, you cannot ask for more. But, I object, how can a statement exclusively about the past also be a statement regarding rational belief with respect to future occurrences? What is the force of the term "rational" here? How is it defined? He takes me back to the beginning and explains the weighting of the state descriptions and the definition of "degree of confirmation." This is what we *mean* by "rational belief"; it is *defined* in terms of "degree of confirmation," which is a logical relation between evidence and hypothesis. My dissatisfaction remains. Will I be right in a high percentage of cases, I ask, if I expect occurrences that are highly probable? Will I be right less often if I expect occurrences that are quite improbable? If I bet in terms of rational degree of belief, will I win my fair share of bets? I cannot say for sure that you will, he replies, but that is a reasonable sort of thing to expect. But, I rejoin, in a sense of "reasonable" that has *no demonstrable connection whatever* with what actually happens in the future! As a matter of fact, you have chosen one from an infinitude of possible definitions of "degree of confirmation" that could have been used to define "rational belief." If you had chosen otherwise, what you now call "rational" might be considered highly irrational, and conversely. In the presence of such a wide variety of choice of possible definitions of "rational," it makes very good sense to ask, "Why be rational?" The answer, "Because it is rational," will not do, for as we have seen, that answer is equivocal.[99] The choice of a definition of "degree of confirmation" seems entirely arbitrary. Is there any way of justifying the choice? By this time the race is over, and neither of us noticed which horse won. So ends the dialogue.

Carnap's most recent view concerning the justification of the choice of a confirmation function rests the matter on our inductive intuition.[100] He has enunciated a number of axioms, and for each he offers an intuitive justification. These axioms determine the limits of the class of acceptable confirmation functions. It is no easy matter to articulate our inductive intuitions in a clear and consistent way. Intuitions are notoriously vague, and they tend to conflict with each other. By dint of careful reflection and comparison, as well as by consideration of a wide variety of examples of inductive reasoning, Carnap believes we can arrive at a fairly precise characterization of rational belief. This is the task of inductive logic, or the theory of logical probability.

Carnap's answer to the question of justification puts him, I think, rather close to those who adopt a postulational approach to the problem of induction, and to those who espouse an "ordinary language" dissolution of the problem. Perhaps the search for plausible postulates is nothing other than Carnap's search for intuitively justifiable axioms. At the same time, the search for axioms that characterize what we intuitively recognize as reasonable inductive inference would presumably come close to capturing the fundamental ordinary meaning of "rational belief." I have already indicated my reasons for dissatisfaction with both of these approaches. I think these reasons apply with the same force to Carnap's theory of degree of confirmation, as well as to any other version of the logical theory.

The logical theory of probability has no difficulty in meeting the criterion of *admissibility*.[101] Carnap has laid great stress upon the importance of avoiding incoherent betting systems, so he has been careful to insist upon an explication that conforms to the axioms of the mathematical calculus. Moreover, since degree-of-confirmation statements are analytic, they can, in principle, be established by mathematical computation. The logical interpretation has no difficulty in meeting the criterion of *ascertainability*. It is the criterion of *applicability* that poses the greatest difficulty for the logical interpretation. This theory provides no reason for supposing any connection whatever between what is probable and what happens often. It seems to provide no basis for expecting the probable in preference to the improbable. In my opinion, it lacks predictive content and thus fails to qualify as "a guide of life." This difficulty concerning the criterion of applicability is Hume's problem of induction once again: Can we offer any justification for the expectations we form, on the basis of observation and inductive inference, concerning unobserved matters of fact?

4. The Personalistic Interpretation. There is another outgrowth of the crude subjectivistic interpretation which, like the logical interpretation, satisfies the criterion of admissibility. It, too, substitutes the concept of rational belief for the concept of actual belief, but in a different way.[102] The purely subjective interpretation ran into trouble, we recall, not because of any individual assignment of probability values, but because of an inadmissible combination. This fact reveals a general feature of the probability calculus. Except for a few essentially vacuous cases, the probability calculus does not, by itself, establish any probability values. It does enable us to compute some values after some others have been supplied. We may thus distinguish fundamental probabilities from derived probabilities, although there is nothing absolute about this distinction. In one context a particular probability may be fundamental; in another, that same probability may be derived. The point of the distinction is to emphasize the fact that some fundamental probabilities are required as a basis for mathematical derivation of other probabilities. The probability calculus provides the relation of fundamental to derived probabilities, but it does not furnish the fundamental probabilities.

The personalistic theory allows, in effect, that the fundamental probabilities are purely subjective degrees of actual belief, but the probability calculus sets forth relations among degrees of belief which must be satisfied if these degrees are to constitute a rational system. Although the fundamental probabilities are subjective, their relation to derived probabilities is objective. There is a parallel with deductive logic. If I believe that all men are mortal, that Socrates is a man, and that

Socrates is immortal, I am guilty of irrationality, in the form of direct logical contradiction, in my system of beliefs. These three beliefs are not jointly tenable. Logic does not tell me which of them is incorrect, or which I ought to revise, but it does tell me that some revision is in order. Similarly, if my degree of belief in each possible outcome of a toss of two dice is equal to my degree of belief in every other outcome, if this degree of belief is unaffected by the outcomes of previous tosses, and if my degree of belief in getting double six at least once in twenty-four tosses exceeds my degree of belief in not getting it, then once again my system of beliefs is irrational. I have not become involved in a logical contradiction, unless I have already made some explicit commitment to the probability calculus for the ordering of beliefs, but there is another type of irrationality. The probability calculus tells me that. It does not determine which degree of belief is to be altered, but it does tell me that some alteration is required. I have already indicated the nature of the irrationality exhibited by degrees of belief that violate the relations stipulated by the probability calculus. Such systems of beliefs give rise to incoherent betting systems—i.e., to systems of bets such that the individual *must lose* no matter what the outcome of the occurrences upon which the wagers are made. An individual with degrees of belief of this sort will be willing to accept a series of bets in which book is made against him. A bookie who knew his beliefs could be certain of making him lose. The desire to avoid incoherent betting systems is an expression of a very practical type of rationality.

Since the personalistic theory of probability condemns as irrational any system of beliefs that violates the probability calculus, and admits as probabilities only systems of beliefs that conform to the calculus, it automatically satisfies the *admissibility* criterion. It also satisfies the *ascertainability* criterion, I believe. Any theory that pretends to treat degrees of belief in a scientific manner must have some way of measuring them. One can, of course, simply ask people how strongly they believe a given proposition and accept their verbal answer as to the degree of their belief. This method is not especially satisfactory. We seem to get more reliable information about subjective states by watching how people behave. Although the actual measurement of degrees of belief involves certain technical complications, the basic idea is rather straightforward.[103] Suppose, to use an example of L. J. Savage, that a subject *S* stands with an egg in each hand—a white egg in one hand and a brown egg in the other. We want to know whether he believes more strongly that the white egg is rotten or that the brown one is. We offer him a choice between two alternatives: either we will give him $1 if the white egg is rotten but nothing if it is not, or we will give him $1 if the brown egg is rotten but nothing if it is not. We promise, moreover, to replace both eggs with eggs of guaranteed freshness. By seeing which alternative S chooses we see which proposition he believes more strongly. In similar fashion, we can offer him alternatives with different sums of money, say a $1.50 if the brown egg proves rotten but still only $1 if the white egg does. We can, moreover, make comparisons with other kinds of situations. For example, we can offer him a choice between $2.50 if the white egg proves rotten and a $2 ticket on Kentucky Kid to win in the next race. By presenting a large number of choices of the foregoing kinds, a great deal can be learned about the subjective probabilities of S. Interesting techniques for measuring degrees of belief have been developed, and there is no reason to doubt that they can be further refined and perfected. It would be distinctly unwarranted to suppose that degrees of belief, because they are subjective, do not admit of objective empirical investigation. The measurement of degrees of belief comes down, fundamentally, to a determination of the kinds of bets a person is willing to accept. There is, consequently, an intimate connection between subjective probabilities and betting behavior, and this accounts for the fact that a strong emphasis upon the coherence of betting systems and the criterion of admissibility is entirely appropriate to the personalistic interpretation.

The personalistic theorists espouse a viewpoint that demands great tolerance concerning probabilities. They maintain, as would many theorists of other persuasions, that conformity of degrees of belief with the probability calculus is a necessary condition of rationality. Most other theorists would deny that it is a sufficient condition as well. If, however, the requirement of coherence is all that is necessary for a set of beliefs to be rational, it is possible to be rational and to hold beliefs that are incredibly irrational according to any of the usual standards. You are allowed any belief you can mention, as long as it is logically consistent, about unobserved events. You cannot be convicted of irrationality as long as you are willing to make the appropriate adjustments elsewhere in your system of beliefs. You can believe to degree 0.99 that the sun will *not* rise tomorrow. You can believe with equal conviction that hens will lay billiard balls. You can maintain with virtual certainty that a coin that has consistently come up heads three quarters of the time in a hundred million trials is heavily biased for *tails!* There is no end to the plain absurdities that qualify as rational. It is not that the theory demands the acceptance of such foolishness, but it does tolerate it.

It should be evident that the difficulties being mentioned fall under the criterion of *applicability*. There is, I think, very serious trouble here. Given any observations whatever of past occurrences, the probability calculus does not, by itself, determine what we must believe about the future, nor does it determine the strengths of any such beliefs.[104] For instance, a coin that has been observed to come up heads one hundred million times in a row may still be a fair coin that comes up tails about half the time in the long run. There is no logical contradiction in this supposition, and the probability calculus cannot fault it. The personalistic theory therefore leaves entirely unanswered our questions about inductive inference. It tolerates *any kind* of inference from the observed to the unobserved. This amounts to an abdication of probability from the role of "a guide of life."

Personalistic theorists have placed great emphasis upon an aspect of their theory I have purposely suppressed until now. I have wanted to stress the essential emptiness of their official concept of rationality. They have, however, explained at length how reasonable men can arrive at a high level of agreement on matters regarding which they originally disagreed, if they will only consider the evidence that is available to all. The mechanism by which this kind of consensus emerges is inference according to Bayes' theorem. All that is necessary is that we all make certain minimal concessions with respect to prior probabilities. I shall discuss Bayes' theorem, prior probabilities, and their roles in scientific inference at a later stage.[105] At present I simply want to remark that the concept of rationality involved in being a "reasonable man" is somewhat different from the notion of rationality defined solely in terms of a conformity of subjective probabilities to the mathematical calculus. A tiny degree, at least, of open-mindedness is now a requirement. This seems to me indicative of a general feature of the personalistic approach. It has correctly identified certain necessary conditions of rationality for subjective probabilities, but additional conditions must be found before we can pretend to have a viable conception of rationality and a concept of probability that satisfies the applicability criterion. It is essential that the additional conditions be stated explicitly and justified. This is precisely where Hume's problem of induction lies for the personalistic theory.

It is noteworthy that Carnap has found his logical interpretation of probability growing closer to the personalistic conception in the last few years.[106] He applauds the coherence requirement and the justification for it. He differs from the personalists, however, in his insistence upon many additional axioms beyond those of the mathematical calculus itself. In all of this I am in complete agreement with Carnap's evaluation of the situation, although I do regard his intuitive justification of his additional axioms as insufficient.

5. *The Frequency Interpretation.* According to a leading version of the frequency interpretation, probability is defined in terms of the limit of the relative frequency of the occurrence of an attribute in an infinite sequence of events.[107] To say, for instance, that the probability of getting heads with this coin is one half means that, in the potentially infinite sequence of tosses of the coin in question, the relative frequency with which heads occurs converges to the value one half. We may think of three coordinated sequences: the sequence of flips F of the coin, the sequence of results (heads H or tails T), and the sequence of fractions representing the relative frequency of heads up to and including that place in the sequence. Here are the results for the initial section of an actual sequence of tosses:

$$\begin{matrix} F & F & F & F & F & F & F & F & F & F & \ldots \\ H & T & H & H & H & T & H & T & H & H & \ldots \\ 1/1 & 1/2 & 2/3 & 3/4 & 4/5 & 4/6 & 5/7 & 5/8 & 6/9 & 7/10 & \ldots \end{matrix}$$

In the third sequence, the denominator of each fraction represents the number of flips in the first sequence up to and including that place, while the numerator represents the number of heads in the second sequence up to and including that place. If the probability of heads is one half, according to the frequency interpretation, the fractions in the third unending sequence converge to the limit one half. The term "limit" is used in its standard mathematical sense:

> The sequence f_n ($n = 1, 2, 3, \ldots$) has the limit L as n goes to infinity if and only if, for every positive ϵ, no matter how small, there exists a number N such that, if $n > N$, $|f_n - L| < \epsilon$.

This definition means, as it is sometimes informally put, that the relative frequencies become and remain as close to L as you like for sufficiently large numbers of elements of the sequence.

In mathematics, the foregoing definition is usually applied to sequences that are generated according to some mathematical rule, for example, $1/n$ or $1/2^n$ where n runs through the positive integers. Each of these sequences has, of course, the limit zero. The sequence of coin tosses and the associated sequence of relative frequencies are not given by a mathematical rule; instead, they are generated by a set of empirically given physical events. I leave open the question of whether there is, in fact, a mathematical rule that would generate precisely the same sequence. If there is, we surely do not know what it is. Some authors have raised doubts about the meaningfulness of talking about limits in sequences of this sort, because there is no known or knowable rule from which to deduce whether or not the sequence has a limit. Nevertheless, the terms in which the definition is framed are entirely meaningful in application to physical events: "every," "there exists," "less than," "difference," "positive number," etc. They are combined, it is true, into an expression that cannot be verified deductively or mathematically, but this should be no source of concern. We are dealing with induction, not deduction. It remains to be seen whether we may properly speak of *inductive verification* of statements about limits of relative frequencies in empirically given sequences.[108]

Assuming that the concept of the limit can be tolerated, it is rather easy to show that the frequency interpretation satisfies the criterion of *admissibility.*[109] Furthermore, there is at least one fundamental and important sense in which the frequency interpretation satisfies the criterion of *applicability.* A statement about the probability of a particular type of event is an objective statement about the frequency with which events of that type will occur. Such statements are synthetic, and they have predictive content by virtue of applying to future events. I do not mean to say that the frequency interpretation faces no difficulties connected with applicability; there are, in fact, serious problems of this kind which I have no intention of minimizing. I shall discuss them later. First, however, we should turn our attention to the problem of ascertainability, for this is the crucial problem facing the frequency theory.

When a sequence is generated by a known mathematical rule, we can, as already noted, deduce statements about limits. We are not dealing with such cases. When a sequence is generated by a physical process that is well understood in terms of accepted physical theory, we may be able to make theoretical inferences concerning convergence properties. For instance, our present knowledge of mechanics enables us to infer the frequency behavior of many kinds of gambling mechanisms.[110] Our theory of probability must allow room for inferences of this kind. The basic probem, however, concerns sequences of events for which we are lacking such physical knowledge. We are dealing with the problem of induction, so we must not assume large parts of inductive science. Instead, we shall consider the question of what inferences, if any, concerning the limit of the relative frequency in a sequence can be made solely on the basis of observations of the initial portions of such sequences. Any such initial section contains, of course, a finite number of members. The first point to emphasize is, as already noted, that we can *deduce* no statement about the limit from a description of any initial section. *Any* value of the relative frequency in an observed initial section of *any* length is compatible with *any* value for the limit. This rather obvious fact has sometimes been taken to destroy the whole enterprise of attempting to make inferences concerning limits of relative frequencies. No such interpretation is warranted. All we can properly conclude is that the problem we are facing is an inductive problem, not a deductive problem, so it cannot have a deductive answer. We are dealing with ampliative inference; the inference from an observed relative frequency to the limit of the relative frequency is certainly of this type.

The situation is, actually, even worse. There is no guarantee that the relative frequencies converge to any limit at all. It is possible to define sequences of events for which the relative frequencies do not converge, and we cannot be sure that such sequences do not occur in nature. This fact should remind us once again of Hume's problem of induction. If a sequence of occurrences does manifest limiting frequencies, it exhibits a type of uniformity—a statistical uniformity. We cannot know a priori that nature is uniform, as Hume showed, and this result applies as much to statistical uniformities as to any other kind.

Hans Reichenbach, a leading proponent of the frequency interpretation, was well aware of all these difficulties, and he appreciated the force of Hume's arguments as applied to the problem of inferring limits of relative frequencies. He maintained, nevertheless, that there is an inductive rule for inferring limits of relative frequencies from observed frequencies in finite initial sections, and this rule is amenable to justification. The justification is a pragmatic one, and it constitutes a refinement of the pragmatic justification discussed above. Even in the refined form, the justification is not adequate, as I shall shortly explain, but I think it may constitute an indispensable part of an adequate justification.

Suppose we have an infinite sequence of events A, and we want to infer the limit of the relative frequency with which these events exhibit the characteristic B. Let $F^n(A, B)$ be the relative frequency of B among the first n members of A. We seek a rule for inferring from the value of $F^n(A, B)$ in an observed sample of n members to $\lim F^n(A, B)$ as $n \rightarrow \infty$. Reichenbach offers the rule of induction by enumeration for this type of inference:

Rule of Induction by Enumeration: Given that $F^n(A, B) = m/n$, to infer that $\lim_{n \to \infty} F^n(A, B) = m/n$.

It is to be understood that the initial section of n members whose relative frequency is given comprises all observed instances of A; when more instances of A are observed a new inference is made from the observed frequency in the larger sample, and the earlier inference from the smaller sample is superseded. The rule of induction by enumeration provides, in short, for the projection of observed statistical uniformities into the unobserved future. It lets us infer that the observed frequency is the same as the limit.[111]

Reichenbach's justification proceeds by acknowledging that there are two possibilities, namely, that the relative frequency with which we are concerned approaches a limit or that it does not, and we do not know which of these possibilities obtains. If the limit does exist, then so does the probability $P(A, B)$; if the limit does not exist, there is no such probability. We must consider each case. Suppose first that the sequence $F^n(A, B)$ $(n = 1, 2, 3, \ldots)$ has no limit. In this case any attempt to infer the value of that (nonexistent) limit is bound to fail, whether it be by induction by enumeration or by any other method. In this case, all methods are on a par: they are useless. Suppose, now, that the sequence does have a limit. Let us apply the rule of induction by enumeration and infer that the observed frequency matches the limit of the relative frequency to whatever degree of approximation we desire. We persist in the use of this rule for larger and larger observed initial parts of our sequence as we observe larger numbers of members. It follows directly from the limit concept that, for any desired degree of accuracy whatever, there is some point in the sequence beyond which the inferred values will always match the actual limit within that degree of approximation. To be sure, we cannot say beforehand just how large our samples must be to realize this condition, nor can we be sure when we have reached such a point, but we can be sure that such exists. There is a sense, consequently, in which we have everything to gain and nothing to lose by following this inductive procedure for ascertaining probabilities—i.e., for inferring limits of relative frequencies. If the probability whose value we are trying to ascertain actually exists, our inductive procedure will ascertain it. If the probability does not exist, we have lost nothing by adopting that inductive procedure, for no other method could have been successful in ascertaining the value of a nonexistent probability.[112]

It is worth comparing the foregoing justification of induction by enumeration for the ascertainment of values of probabilities with the earlier and much looser pragmatic justification of induction. In the earlier argument, we considered the two possibilities that nature is uniform or that nature is not uniform. At the time, I remarked that these two alternatives are not necessarily exhaustive, and that we needed to be much clearer about the extent and degree of uniformity required. In the latter argument, we deal with the same set of alternatives, but now we can specify exactly what the uniformity consists in—namely, that the sequence of relative frequencies whose limit we are trying to infer is convergent. Moreover, the alternatives considered are exclusive and exhaustive: The sequence has a limit or it does not. In the earlier argument, it was not clear what constituted an inductive method; in the latter argument, it is specified precisely as the rule of induction by enumeration. In the earlier argument, we used notions like success and failure of inductive methods, again without any precise indication of their meanings. In the latter argument, their force is clear. To begin with, we choose some degree of approximation ϵ that will satisfy us for this particular probability. Then, success consists in inferring the value of the limit of the relative frequency on the basis of a sample large enough that this inference, as well as all subsequent inferences on the basis of larger samples, are accurate within $\pm \epsilon$. When Reichenbach claimed, in connection with the looser version of his argument, that the inductive method will work if any method will work, some people thought he was actually making an inductive generalization. In the tighter version of the argument we see that this claim is not inductive in character. It is a rather obvious—but nonetheless important—analytic consequence of the definitions of "limit of a sequence" and "induction by enumeration." It is not a surreptitious inductive justification of induction. The justification is deductive, but not in the way Hume's arguments ruled out. It does not prove deductively that all or most inductive inferences with true premises will have true conclusions. It does prove that induction by enumeration will enable us to make accurate inferences concerning limiting frequencies if any method can. This proof is, I submit, valid.

The chief defect in Reichenbach's justification is that it fails to justify a unique inductive rule, but rather, it justifies an infinite class of inductive rules equally well. Although induction by enumeration will work if any method will, it is *not* the only rule that has this characteristic. Reichenbach's argument shows that the same can be said for an infinite class of inductive rules. Reichenbach was aware of this fact, and he characterized the class as *asymptotic.*[113] A rule is asymptotic if it shares with the rule of induction by enumeration the property of yielding inferences that converge to the limit of the relative frequency whenever such a limit exists. The rule of induction by enumeration takes the observed frequency itself as the inferred value of the limit of the relative frequency. Any other rule that gives inferred values differing from the observed frequency, but in such a way that the difference converges to zero as the sample size increases, is asymptotic. Let $I^n_R(A, B)$ be the inferred value, on the basis of a sample of size n, of the limit of $F^n(A, B)$ according to rule R. (When it is clear what rule is involved, the subscript R may be omitted.) A rule is asymptotic if and only if

$$I^n(A, B) - F^n(A, B) \rightarrow 0 \text{ as } n \rightarrow \infty$$

Notice that this definition does not depend upon an assumption that the limit of the relative frequency exists, for the differences may converge whether or not the relative frequencies do. Notice also that induction by enumeration is asymptotic, for in its case the above difference is identically zero. It is evident that every asymptotic rule has the property used to justify induction by enumeration: If the relative frequency has a limit, the rule will produce inferred values that become and remain accurate within any desired degree of accuracy.

The fact that there are infinitely many asymptotic rules is not, by itself, cause for dissatisfaction with Reichenbach's argument. If this infinity contained only a narrow spectrum of rules that yield similar results and that quickly converge to one another, we could accept the small degree of residual arbitrariness with equanimity. The actual case is, however, quite the opposite. The class of asymptotic rules is so broad that it admits *complete* arbitrariness of inference. Although all of the rules of this type give results that converge to one another, the convergence is nonuniform. This means, in effect, that for any finite amount of evidence we might have, they are actually completely divergent. Given a sample of *any* finite size, and given *any* observed frequency in that sample, you may select *any* number from zero to one inclusive, and there is some asymptotic rule to sanction the inference that this arbitrarily chosen number is the probability. The class of asymptotic rules, taken as a whole, tolerates any inference whatever regarding the limit of the relative frequency.

For any particular probability sequence, certain asymptotic rules will give results that converge to the accurate value more rapidly than will others of the asymptotic rules. It might be tempting to suppose that induction by enumeration would give quicker convergence to the correct value than do the other asymptotic rules, or at least that it will usually do so, but no such thing can be proved. We have to find some other ground for choice if we hope to justify the selection of a unique rule. Reichenbach attempted to argue for induction by enumeration on the grounds of "descriptive simplicity," but this argument seems to me patently inapplicable.[114] Descriptive simplicity can be invoked only in the case of theories, statements, or rules that are empirically equivalent.[115] Although the asymptotic rules do all converge "in the limit," they cannot possibly be regarded as empirically equivalent because of the complete arbitrariness of inference they tolerate as a class.[116]

We must conclude that Reichenbach did not succeed in showing that the frequency interpretation can meet the criterion of *ascertainability*. Finding a way to satisfy this criterion is the form in which Hume's problem of induction arises for the frequency interpretation. In spite of Reichenbach's failure to give an adequate solution to this problem, he did accomplish two things. First, he showed us a direction in which to look for an answer to Hume's problem. The fact that his particular attempt at a pragmatic justification did not succeed does not prove that no pragmatic justification is possible. Second, he gave a convincing argument, I believe, for the rejection of nonasymptotic rules of inference. Although not many authors would accept Reichenbach's reason for insisting upon the asymptotic character of rules of inference, a large number and wide variety of theorists do insist upon the condition of asymptoticity, or *convergence* as I shall call it.[117] For instance, Carnap adopts just this condition as one of his axioms. The personalistic theorists achieve a similar result through the use of Bayes' theorem with certain requirements placed upon the prior probabilities. In spite of the fact that it is not often used for this purpose, Reichenbach's argument is a consideration in favor of each of these concepts of probability insofar as they deal with the problem of inferring "long-run" frequencies. It seems widely recognized, explicitly or implicitly, that it would be foolish indeed to apply a method for inferring limits of relative frequencies such that, if the limit exists, persistent use of the method is bound to produce ever-recurrent error. Reichenbach's argument is at least a beginning toward a solution of the problem of ascertainability for the frequency interpretation, but it is also a beginning toward a solution of the problem of applicability for the logical and personalistic interpretations. The crucial question is whether we can find additional grounds by which to narrow significantly the class of acceptable inductive rules.

Although I am convinced that the problem of ascertainability is the most fundamental difficulty the frequency interpretation faces, there are, as I said earlier, certain important problems of applicability. When I presented the criterion of applicability, I enumerated several different ways in which the concept of probability functions. It has often been maintained that the frequency interpretation, while adequate to fulfill some of these functions, is patently inadequate to others. Carnap, for example, does not advocate a logical interpretation to the complete exclusion of all other interpretations. Instead, he *insists* upon two concepts of probability, the logical concept and the frequency concept. The frequency concept may be entirely appropriate for the interpretation of statistical laws of physics, for example, but it is unsuitable as a basis for explicating the concept of probability as it enters into the confirmation of scientific hypotheses by observational evidence. For this purpose, he claims, logical probability is needed. I shall return to this topic below, for the problem of confirmation must be considered in detail. If Carnap is right, the frequency interpretation alone cannot satisfy the criterion of applicability, but it can in conjunction with other interpretations.

The frequency interpretation also encounters applicability problems in dealing with the use of probability as a guide to such practical action as betting. We bet on *single* occurrences: a horse race, a toss of the dice, a flip of the coin, a spin of the roulette wheel. The probability of a given outcome determines what constitutes a reasonable bet. According to the frequency interpretation's official definition, however, the probability concept is meaningful only in relation to infinite sequences of events, not in relation to single events. The frequency interpretation, it is often said, fails on this extremely important aspect of the criterion of applicability.

Frequentists from Venn to Reichenbach have attempted to show how the frequency concept can be made to apply to the single case. According to Reichenbach, the probability concept is extended by giving probability a "fictitious" meaning in reference to single events. We find the probability associated with an infinite sequence and transfer that value to a given single member of it.[118] For instance, we say that the probability of heads on any particular toss of our coin is one half. This procedure, which seems natural in the case of the coin toss, does involve basic difficulties. The whole trouble is that a given single event belongs to many sequences, and the probabilities associated with the different sequences may differ considerably. The problem is to decide from which sequence to take the probability that is to be attached "fictitiously" to the single event.

According to the frequency interpretation, probability is a relation between two classes. The notation, "$P(A, B)$," reflects this fact by incorporating expressions for two classes, one before the comma and one after it. The class mentioned first is the *reference class;* the other is the *attribute class.* In dealing with the problem of the single case, the attribute class gives us no particular trouble. The terms of the bet determine which attribute we seek: double six, heads, the ace of spades, etc. The problem of the single case *is* the problem of selecting the appropriate reference class. Consider, for example, how to determine what premium a given individual should be charged for his automobile insurance. The insurance company tries to assign him to a category of drivers who are similar to him in relevant respects. It matters, for instance, whether the driver is male or female, married or unmarried, an urban or a rural dweller, a teenager or not, etc. It does not matter what color the car is or whether it has an odd or even license number. Reichenbach said that one should choose the narrowest reference class for which reliable statistics are available. I would say, instead, that the single case should be referred to the *broadest homogeneous reference class* of which it is a member. In either formulation, the intent is fairly straightforward. A probability is something that has to be established inductively, and in order to ascertain the probability we must have enough instances to be able to make an inductive generalization. Thus, we do not want to try to refer single cases to classes that are too narrow, for if we do we will not have enough evidence upon which to base our inference. At the same time, we want our reference class to contain other relevant cases, not irrelevant ones. *Statistical relevance* is the key concept here. Suppose we ask for the probability that a given individual x has a characteristic B. We know that x belongs to a reference class A in which the limit of the relative frequency of B is p. If we can find a property C in terms of which the reference class A can be split into two parts $A \cap C$ and $A \cap \bar{C}$, such that

$$P(A \cap C, B) \neq P(A, B)$$

then C is *statistically relevant* to the occurrence of B within A. Of course, C must be the sort of property whose occurrence in an individual can be detected without knowing whether that particular entity also has the property B. If there is no such property C by means of which to effect a relevant subdivision of A with respect to the occurrence of B, A is *homogeneous* with respect to B. Consider the reference class A of tosses of our coin with respect to the attribute B of heads. Let us suppose we know that the probability $P(A, B) = 1/2$. We do not have any way of making a relevant subdivision in this reference class. Let C be the property of being an even toss, so $A \cap C$ consists of the tosses numbered 2, 4, 6, ... Since the coin is fair, $P(A \cap C, B) = P(A, B) = \frac{1}{2}$. The same situation obtains if C is the class of tosses immediately following tosses on which heads occurred, if C is the class of tosses made before dark, or if C is the class of tosses made on Monday, Wednesday, or Friday. A homogeneous reference class is the essence of a fair gambling mechanism, as Richard von Mises has carefully explained.[119] By contrast, let A be the class of licensed drivers in the United States. Let B be the class of drivers who have accidents involving over $50 damage to the car they are driving. We all know that A is not homogeneous with respect to B in this case. A can be subdivided in terms of a number of relevant characteristics such as age, sex, habitat, amount of driving done, etc. These subdivisions must be made before trying to assign a probability to the occurrence of an accident for a particular driver x.

It would be most unrealistic to suppose that we can fulfill the requirement of selecting the broadest homogeneous reference class in all cases in which we have to make practical decisions about single events. We may suspect that a given reference class is inhomogeneous, but not know of any way to make a relevant partition of it. Under these circumstances let us say that the class is *epistemically homogeneous;* reference classes of this type are the best we have in many cases until knowledge is further advanced. Sometimes we know that a reference class is inhomogeneous, but it would simply be impractical to carry out a relevant subdivision. The coin example is a paradigm. With elaborate enough measurements of the initial conditions of the toss, and with a fancy enough set of calculations, we could, in principle, do a pretty good job of predicting which side will come up on a given toss. It is not worth the effort, and besides, the rules of the game do not allow it. Under these circumstances, let us say that the reference class is *practically homogeneous.* We often make do with reference classes of this kind. Moreover, a given reference class may admit of a relevant subdivision, but the alteration of probability achieved thereby may be too small to be worthwhile. The relevant subdivision might also effect a serious reduction in available statistics. The choice of a reference class is an extremely practical affair, in which we must balance a number of factors such as the size of the class, the amount of statistical evidence available, the cost involved in getting more data, the difficulty in effecting a relevant subdivision, and the amount at stake in the wager or decision.

Carnap has made an extremely important and useful distinction between inductive logic proper and the methodology of induction.[120] I alluded briefly to this distinction in explaining his view that inductive logic does not contain rules for the acceptance of hypotheses. Inductive logic contains the systematic explication of degree of confirmation, and analytic degree of confirmation statements that hold for a given confirmation function in a given language. The methodology of induction contains rules for the application of inductive logic—that is, rules that tell us how to make use of the statements of degree of confirmation in deciding courses of practical action. As I indicated, the requirement of total evidence is one of the important methodological rules, and the rule of maximizing estimated utility is another.[121] These rules tell us how to use the results of inductive logic, but they do *not* provide for inductive inferences in which the conclusion can be detached and asserted if the premises are true and the degree of confirmation is sufficiently high.

Reichenbach, unfortunately, did not make the same clear distinction between probability theory proper and the practical rules for the application thereof. Such a distinction would have been helpful, particularly for the problem of the single case. Reichenbach admits that the meaning of "probability" for the single case is fictitious. He does not offer rules which enable us to assign probability values univocally. It is apparent that very practical considerations determine the value actually assigned. It would have been better, I think, if he had refused to apply the term "probability" to single events at all, but had instead reserved some other term such as "weight" which he often used for this purpose.[122] We could then say that probability is literally and exclusively a concept that applies to infinite sequences, not to single cases. If we want to find out how to behave regarding a single case, we must use probability knowledge; the problem is one of deciding how to apply such knowledge to single events. Our rules of application would tell us to find an appropriate reference class to which our single case belongs and use the value of the *probability* in that infinite sequence as the *weight* attached to the single case. The rule of selecting the broadest homogeneous reference class becomes a rule of application, not a rule for establishing values of probabilities within the theory proper. This approach shows why very down-to-earth practical considerations play an important role in determining *weights* of single events. Weights can be used to determine betting odds, to compute mathematical expectations for various programs of action, and for other equally practical purposes. This treatment of the single case shows a deep analogy between the requirement of total evidence and the rule for selecting the appropriate reference class. Both rules are part of methodology, not of probability theory. Each rule requires us to make use of all relevant available information in arriving at practical decisions.

The present treatment of the single case should also remove any temptation to suppose that statements about single cases can be detached and asserted independent of evidence and without reference to the weight. The methodological rules for handling single events are not rules of inference. There is an important reason for this. Suppose, as most authors on induction have, that we can assert hypotheses about single events if we have sufficient inductive evidence. P. F. Strawson gives an apt example: "He's been travelling for twenty-four hours, so he'll be very tired." [123] Strawson remarks, "Plainly the statement made by the first clause . . . is regarded as a reason for accepting the statement made by the second clause. The second statement . . . is in some sense a *conclusion* from the first. . . ." [124] Accepting the conclusion that "he'll be very tired" may be well and good in ordinary circumstances, but it would be foolhardy if the question is whether to engage him in mortal combat as he steps from the train. Just how we regard the "conclusion" depends upon what is at stake. The situation is made quite clear by the famous lottery paradox. Imagine a fair lottery with as large a number n of tickets as you like. The probability of a given ticket winning is $1/n$, which is as small as you like. The probability that a given ticket will not win is $1 - 1/n$, which is as close to one as you like. If there is some probability value p, such that a hypothesis can be accepted if its probability exceeds p, then we can make the probability that the ticket will not win greater than p. Hence, we can accept the hypothesis. But, the same holds for each ticket in the lottery, so we can deduce that no ticket will win. This contradicts the assumption that it was a fair lottery in the first place. To avoid this sort of difficulty, Carnap does not allow the hypothesis to be detached and asserted. Instead, he permits the use of the value $1/n$ in determining a fair price to pay for a ticket. When probability is used in this way, no paradox arises. I am proposing the same approach as a modification of Reichenbach's treatment of the single case. The weight can be used to determine the price to pay for a ticket, but a large weight does not warrant an assertion about the occurrence or nonoccurrence of a single event.

Carnap's inductive logic, as I have already explained, does not give rise to the assertion of any synthetic statement. Evidence statements are, as noted, ordinarily synthetic, but they are presumably asserted on the basis of observation, not inductive inference. I have explained at length my reasons for objecting to the exclusively analytic character of probability statements. In the frequency interpretation, as I am treating it, synthetic probability statements are asserted on the basis of inductive evidence. This approach allows the use of induction by enumeration, or some other asymptotic method, to infer the limit of the relative frequency—i.e., the probability—from the observed frequency in a finite sample. These *synthetic* probability statements are then applied, according to the methodological rules, in order to obtain weights that can be used for practical decisions. Although the lottery paradox arises if we try to make assertions about the occurrence or nonoccurrence of single events, no such paradoxes can arise in connection with assertions concerning limits of relative frequencies. You can never settle a bet on such assertion. Notice that no probability is assigned to the statement of the limit of the relative frequency. It is simply asserted, on inductive evidence, until additional evidence requires its revision. When that happens, the former value is discarded and replaced by a new value based upon a larger body of inductive evidence.

The fundamental objection I raised against the logical conception of probability, with its analytic degree of confirmation statements, is that such probability statements cannot meet the applicability criterion and cannot function as "a guide of life." The frequency treatment of the single case, as just outlined, does seem to me to meet the applicability requirement.

In a penetrating analysis of the logical interpretation of probability, A. J. Ayer has raised a fundamental question about the requirement of total evidence, namely, what makes total evidence any better than any other kind of evidence? [125] The logical interpretation of probability gives us a whole array of degree of confirmation statements for any given hypothesis, with different degrees of confirmation for different bodies of evidence. All of these statements of degree of confirmation are on a par; they are true but they have no content. We are told to pick the number from the degree of confirmation statement that embodies total available evidence, and that is the number to be used in practical decision making. Why? Logically, this degree of confirmation statement is no better or worse than any of the others. Practically, can we show that we will be more successful if we follow this requirement than we would be if we threw out, say, all evidence collected on Sunday? Hume's argument shows that we can prove no such thing. It is perfectly consistent to suppose that the world is governed by a deity who disapproves of

Sunday data collecting, so he punishes those who use such data by making a great many of their predictions go wrong.

The frequency approach does *not* face the same difficulty. *If* we know the long-run probabilities, a certain type of success is assured by following the methodological rules for handling single events. A given individual deals with a great variety of single events. As an aggregate they may not have much in common. They may be drawn from widely different probability sequences. A man bets on the toss of some dice, he decides to carry his umbrella because it looks a bit like rain, he takes out an insurance policy, he plants a garden, he plays poker, he buys some stock, etc. This series of single cases can, however, be regarded as a new probability sequence made up of the single cases the individual deals with. It is demonstrable that he will be successful "in the long run" if he follows the methodological rules laid down. He will win a few and lose a few, but if he so acts that he has a positive expectation of gain in each case, he will come out ahead in the long run.[126] Of course, as Lord Keynes has reminded us, he will also be dead in the long run.[127]

The foregoing survey of five important attempts to provide an interpretation of the probability concept reveals that each one encounters severe difficulties in meeting at least one of the three criteria of adequacy. The classical and subjective interpretations fail on the criterion of admissibility, the frequency interpretation has difficulties with respect to the criterion of ascertainability, and the logical and personalistic interpretations run into troubles on the criterion of applicability. All of these interpretations have been considered separately. It would be natural to ask at this point whether some combination of interpretations, such as Carnap endorses, would mitigate the problem. The answer is, I think, emphatically negative. The amalgamation of two concepts, one of which is unsatisfactory from the standpoint of ascertainability while the other is unsatisfactory from the standpoint of applicability, yields a theory that has problems of both of these counts. It does not seem to produce a theory that overcomes both sorts of difficulties.

VI. Inferring Relative Frequencies

Theorists of various different persuasions agree that relative frequencies are basically germane to probability theory, whether or not they are willing to *define* "probability" in terms of them. Carnap, who is the greatest proponent of the logical interpretation, insists that there are two concepts of probability—the logical concept and the frequency concept.

He argues, moreover, for an intimate relation between relative frequency and degree of confirmation. The use of probability as a fair betting quotient rests upon its relation to the frequency with which various kinds of events occur. In addition, degree of confirmation can be interpreted in appropriate circumstances as an estimate of a relative frequency—namely, when the hypothesis being confirmed is a statement about a single occurrence and the evidence is statistical evidence concerning the frequency of that type of occurrence in observed cases.[128] In these and other matters, Carnap seems not far from F. P. Ramsey, who is properly regarded as the originator of the personalistic interpretation. Ramsey, too, takes great care to point out fundamental relations between frequencies and partial beliefs.[129]

The reason for the fundamental connection is quite plain to see. Suppose a die is about to be tossed twelve times, and suppose that the relative frequency with which the side six will show in this set of tosses is one sixth. This means that six will occur twice. Suppose, moreover, that Smith agrees to bet on six at odds of five to one, and Jones agrees to accept these bets. The result is that Smith wins two bets and collects $5 on each, for a total of $10, while Jones wins ten bets and collects $1 on each, also for a total of $10. The two players come out even, neither winning anything from the other. This game would, of course, be utterly pointless if both players knew for sure what the outcome would be, but they cannot know any such thing in advance. However, each player would do his best to guess or infer the relative frequency of six, for the relative frequency determines what constitutes a fair bet or a bet that is advantageous to either party.

Let us approach the question of justifying a method for inferring relative frequencies as a problem of basic concern to any probability theory. Reichenbach attempts to justify induction by enumeration as the fundamental method for this purpose, but, as we have already noted, there are infinitely many alternative candidates, and his argument does not succeed in showing that induction by enumeration is superior to all the rest. Two alternatives, the a priori method and the counterinductive method, were mentioned above, but since neither is asymptotic they are both unacceptable. I shall assume that the reasons already given are a sufficient basis to insist upon asymptoticity in inductive rules.

The rule of induction by enumeration is the simplest of the asymptotic rules. This does not constitute any kind of justificatory argument, unless we can show that there are good reasons for preferring a simple rule to more complex ones, but it does make that rule a convenient point of departure. If a rule is asymptotic, the values we infer for the limit of the relative frequency must be determined, at least in part, by observed frequencies in observed samples, for the inferred values must converge to the observed frequencies as the sample size increases. For purposes of systematic exploration, it is convenient to represent any asymptotic rule R as follows:

Given $F^n(A, B) = \frac{m}{n}$, to infer $\lim_{n \to \infty} F^n(A, B) = \frac{m}{n} + c$,
where $c \to 0$ as $n \to \infty$.

Induction by enumeration is the rule that results when c is identically zero. The other asymptotic rules impose some "corrective term" c which produces a difference between the observed frequency and the inferred value of the limit. The term c is obviously not a nonzero constant; what it is a function of will be a question of primary interest to us.

There is a basic fact about limits of relative frequencies that will help us to rule out a large class of asymptotic rules. Let A be any reference class, and let $B_1, B_2, \ldots, B_k$ be any mutually exclusive and exhaustive set of attributes in A. This means that every member of A has one and only one of these attributes B_i. For any initial section containing n members of A, let m_i be the number of members of the sample having attribute B_i; m_i/n is the observed frequency of B_i in that sample. For any positive n, all of the corresponding values of m_i must add up to n, and all of the values of the observed frequencies m_i/n must add up to one. Furthermore, no relative frequency can ever be negative. Therefore,

$$\frac{m_i}{n} \geq 0 \text{ and } \sum_{i=1}^{k} \frac{m_i}{n} = 1.$$

The same conditions hold for limits of relative frequencies:

$$\lim_{n \to \infty} F^n(A,B_i) \geq 0 \quad \text{and} \quad \sum_{i=1}^{k} \lim_{n \to \infty} F^n(A,B_i) = 1.$$

I have called these conditions for limits *normalizing conditions*. They are arithmetical truisms, and any statement about limits of relative frequencies that violates them is an outright contradiction. Any rule that produces inferred values of limits that fail to satisfy them is clearly unsatisfactory.[130]

The normalizing conditions would rule out the counterinductive method if it were not already knocked out by the convergence requirement. Suppose we have an urn from which we have drawn a number of balls, and among the observed cases one half are red (B_1), one fourth are yellow (B_2), and one fourth are blue (B_3). Suppose it is given that no other color is exemplified in that urn. The three-color attributes are mutually exclusive and exhaustive. The counterinductive rule requires us to infer that the limit of the relative frequency of B_i is $(n - m_i)/n$ whenever the observed frequency is m_i/n. Thus, we infer that the limit of the relative frequency of red is one half, the limit of the relative frequency of yellow is three fourths, and the limit of the relative frequency of blue is three fourths. These inferred values add up to two, which constitutes a violation of the normalizing conditions and a patent absurdity.

The normalizing conditions must hold for every value of n, so c cannot be a function (not identically zero) of n alone. If it were, we could choose some value of n for which c does not vanish and we would have

$$\sum_{i=1}^{k}\left[\frac{m_i}{n} + c(n)\right] = k \times c(n) + \sum_{i=1}^{k}\frac{m_i}{n} = 1 + k \times c(n)$$

which is a violation of the normalizing conditions.

The normalizing conditions tell us something rather important about the "corrective" term. Since observed frequencies must add up to the same total as the inferred values of the limits, whatever is added to one observed frequency by the "corrective" term must be taken away from other observed frequencies. This feature of the "corrective" term must be built into the rule. When the "corrective" term is seen in this way, it leads us to suspect immediately that the decision as to what observed frequencies should be increased and what ones decreased may be entirely arbitrary.

It is important to note that the "corrective" term cannot be a function of the observed frequency alone. It cannot be a Robin Hood principle that takes from the rich and gives to the poor, nor can it be a biblical principle that gives more to those who already have much and takes away from those who have little. This fact can be proved as follows: Since the normalizing conditions must be satisfied for every value of n, fix n. Consider first a case in which k, the number of attributes, is equal to n, the number of elements in the sample, and in which each attribute occurs just once. In this case,

$$\frac{m_i}{n} = \frac{1}{k} = \frac{1}{n}.$$

According to the normalizing conditions, the inferred values must sum to one, so we have

$$\sum_{i=1}^{k}\left[\frac{m_i}{n}+c\left(\frac{1}{k}\right)\right]=k\times c\left(\frac{1}{k}\right)+\sum_{i=1}^{k}\frac{m_i}{n}=k\times c\left(\frac{1}{k}\right)+1=1.$$

Hence,

$$k\times c\left(\frac{1}{k}\right)=0,$$

and since $k\neq 0$,

$$c\left(\frac{1}{k}\right)=c\left(\frac{1}{n}\right)=0.$$

Now let m_1 be any number such that $0\leq m_1\leq n$. Let $m_i=1$ for $i=2$, $3, \ldots, k$. This time the normalizing condition yields

$$\sum_{i=1}^{k}\left[\frac{m_i}{n}+c\left(\frac{m_i}{n}\right)\right]=\frac{m_1}{n}+c\left(\frac{m_1}{n}\right)+\sum_{i=2}^{k}\left[\frac{m_i}{n}+c\left(\frac{1}{n}\right)\right]=1.$$

Since, as we have just shown, $c(1/n)=0$,

$$\frac{m_1}{n}+c\left(\frac{m_1}{n}\right)+\sum_{i=2}^{k}\frac{m_i}{n}=1=\sum_{i=1}^{k}\frac{m_i}{n}+c\left(\frac{m_1}{n}\right).$$

Since

$$\sum_{i=1}^{k}\frac{m_i}{n}=1,$$

it follows that

$$c\left(\frac{m_1}{n}\right)=0.$$

Since the argument is perfectly symmetrical with respect to i,

$$c\left(\frac{m_i}{n}\right)=0$$

for any possible value of m_i. If c is a function of the relative frequency alone, it is identically zero.

Another possible basis for the "corrective" term is the language in which we are attempting to carry out our inference. For example, the a priori rule given on page 50 makes its inferred value depend upon the number of predicates used to describe the balls drawn from the urn.

Since there are three color *terms* involved, it sanctions the conclusion that the limit of the relative frequency is one third for each. Unlike the counterinductive method, it does not violate the normalizing conditions, but like that rule, it does violate the convergence requirement. This rule, of course, does not really make any use of the observed frequency, but we can still speak of it in terms of the observed frequency plus a "corrective" term. In this case, $c = -m_i/n + 1/k$.

I have argued elsewhere that rules which make the "corrective" term depend upon the features of language are not to be considered acceptable.[131] For example, the inferred value of the limit of the relative frequency of *red* draws from the urn can be changed from one third to one fourth by the simple device of defining two new terms, "dark blue" and "light blue," which are together equivalent to "blue," thereby transforming k from three to four. It seems evident that such an arbitrary alteration of an irrelevant aspect of the language ought not to have any bearing upon the inference we make concerning the limiting frequency of red balls being drawn from the urn.

In order to exclude rules that share this type of defect, I have suggested the *criterion of linguistic invariance* as a requirement to be met by any inductive rule. The general idea behind this criterion is that inductive relations between objective evidence and factual hypotheses depend upon the *content* of the evidence statements and the hypotheses, but not upon the linguistic form in which they are stated. We would all agree, for instance, that it makes no difference whether evidence and hypothesis are stated in terms of the metric system or the English system. They can be stated in either way, and translated back and forth. If an experimental finding strongly confirms a physical hypothesis, it does so regardless of which formulation is chosen. Similarly, given that the evidence and hypothesis can be equivalently formulated in English, French, German, or Russian, the inductive relation is independent of the choice of language. More precisely, the criterion can be stated as follows:

> *The criterion of linguistic invariance:* If S and S′ are statements in the same or different languages such that (1) S asserts that a certain relative frequency $F^n(A, B)$ obtains in a sample of size n, (2) S′ is equivalent to S by virtue of the semantical and syntactical rules of the languages in which they occur, and (3) rule R sanctions the inference from S that
>
> $$\lim_{n \to \infty} F^n(A, B) = p$$

then R must not sanction the inference from S' that

$$\lim_{n \to \infty} F^n(A, B) = q$$

where $p \neq q$.

The foregoing formulation is framed with particular reference to rules for inferring limits of relative frequencies from observed frequencies. More generally, the criterion could be stated:

> If (1) e and e' are any two evidence statements in the same or different languages, (2) h and h' are two hypotheses in the same languages as e and e', respectively, and (3) e is equivalent to e' and h is equivalent to h' by virtue of the semantical and syntactical rules of the languages in which they occur, then the inductive relation between e and h must be the same as the inductive relation between e' and h'.

This general requirement can, it seems to me, be transformed into the following consistency requirement for inductive logic:

> No inductive rule shall permit mutually incompatible conclusions to be drawn from any single consistent body of evidence.

This principle can be illustrated by application to the a priori rule. Given that seven tenths of the observed draws from the urn have yielded red, infer that the limit of the relative frequency of red is one third. This inference (in which the observed frequency enters only vacuously) is formulated in the language containing the predicates "red," "yellow," and "blue." Formulating the parallel inference in the language containing the predicates "red," "yellow," "light blue," and "dark blue," we find we are permitted to infer, from the fact that seven tenths of the observed draws from the urn have yielded red, that the limit of the relative frequency of red is one fourth. These two conclusions from the same consistent body of observational evidence are incompatible. For this reason, as well as failure to satisfy the convergence requirement, the a priori rule is unsatisfactory.[132]

Carnap's earlier systems of inductive logic were subject to precisely the same criticism unless it was blocked by the rather unappetizing requirement of descriptive completeness.[133] Without this requirement, it was possible to show that the addition or removal of irrelevant predicates (not occurring in either the evidence or the hypothesis) could alter the degree of confirmation and consequently the estimate of the relative frequency.

I shall illustrate with a concrete example, but in order to do so I must introduce Carnap's concept of a *Q-predicate*.[134] A Q-predicate is, in certain ways, strongly analogous to a state description. A state description is the strongest type of *statement* that can be formulated in a language for inductive logic; a Q-predicate is the strongest kind of *predicate* that can be formulated in such a language. In our simple illustrative language we had only one predicate "F," so there were only two things we could say about a given individual, namely, that it has the property F, or that it does not have the property F. "F" and "$\sim F$" are the two Q-predicates in that language. Suppose, now, that we expand that language by adding another primitive predicate "G" that is logically independent of "F." In this larger language we can say much more about our individuals, namely, for each individual x there are the following four possibilities:

1. $Fx \cdot Gx$	3. $\sim Fx \cdot Gx$
2. $Fx \cdot \sim Gx$	4. $\sim Fx \cdot \sim Gx$

These are the Q-predicates. Notice that these are not state descriptions, for each one refers to only one individual, while state descriptions must describe each individual. In general, for a language containing k primitive predicates that are logically independent of each other, there are 2^k Q-predicates obtained by either affirming or denying each primitive predicate of any individual. A Q-predicate provides a complete description of any individual. Any predicate in the language is equivalent to one or more of the Q-predicates. For instance, "F" is equivalent to the disjunction of the first two Q-predicates above.

A general formula can be given for the so-called "singular predictive inference" on the basis of the Q-predicates.[135] Let "M" be some predicate —not necessarily a primitive predicate or a Q-predicate—that can be expressed in the language. Let h be the hypothesis that a given unobserved individual has the property M. Let e be the statement that n individuals (not including the one mentioned in h) have been examined, and m of them have been found to exhibit the property M. The probability that an unobserved individual has the property M, on the basis of this evidence, is given by

$$c^*(h, e) = \frac{m + w}{n + 2^k}$$

where w is the number of Q-predicates to which M is reducible, i.e., w is the *logical width* of M. In our miniature language with three individual names and one primitive predicate (two Q-predicates), we calculated the probability that the third individual c had the property F, given that *a* and *b* both had that property. The degree of confirmation was found to be three fourths, in obvious agreement with the above formula, for in that case $n = 2$, $m = 2$, $w = 1$, and $2^k = 2$.

Let us compute the same probability in a new language which is identical to the former language except for the addition of the new primitive predicate "*G*." The evidence *e* and hypothesis *h* remain unchanged. Now, however, as already noted, "*F*" is no longer a *Q*-predicate, but is expressible in terms of the first two *Q*-predicates. Therefore, *w* now equals two and 2^k now equals four. In the new language, as the above formula reveals, the degree of confirmation of precisely the same hypothesis on precisely the same evidence has changed from three fourths to two thirds! This is an obvious violation of the criterion of linguistic invariance.[136]

The criterion of linguistic invariance has, itself, been challenged in various ways. Its strongest defense is, in my opinion, on the grounds that it is a consistency requirement whose violation allows for the possibility of explicit logical contradiction. Suppose, for example, that we can infer from one set of premises S that

$$\lim_{n \to \infty} F^n(A,B) = p$$

and from a logically equivalent set of premises S′ we can infer that

$$\lim_{n \to \infty} F^n(A,B) = q$$

where $p \neq q$. Under the normal rules of logical procedure, we could substitute S for S′ in the second inference, with the result that we would have two distinct values *p* and *q* for the same limit on the basis of one consistent set of premises S. Since it is a fundamental fact that limits of sequences are unique wherever they exist, the foregoing inference has resulted in a self-contradiction.

Moreover, if limits of relative frequencies are to be taken as probabilities, such inferences will lead to a violation of the criterion of admissibility, for the probability calculus requires the uniqueness of probabilities.

The foregoing justification of the criterion of linguistic invariance seems to me to be decisive for any inductive theory that admits rules of acceptance. Carnap's inductive logic admits no such rules, so the criterion of linguistic invariance cannot be defended for its ability to exclude rules of acceptance that lead to contradiction. An inductive logic that has no rules of acceptance has no worries on this score. There is, however, another argument that can be brought to bear. If degrees of confirmation are linguistically variant in the manner illustrated above, they can lead to incoherent betting systems. Consider the odds to be given or taken in a bet on whether the third ball to be drawn from the urn will be red. Since the degree of confirmation of red is (by the second calculation) two thirds, a bettor Smith should be willing to give odds of two to one that the next draw will yield red. At the same time, the degree of confirmation of red is (by the first calculation) three fourths, so Smith should be willing to take odds of three to one that the next draw will be nonred. If he does both of these things, a book can be made against him. A sly bookie could get him to bet $3 on red at odds of one to three, and $1.25 on nonred at odds of two to one. With these bets he must lose whatever happens. If red occurs, he wins $1 on the first bet and loses $1.25 on the second; if nonred occurs, he wins $2.50 on the second bet and loses $3 on the first. This is not a happy situation for Smith.

At one time I thought that the convergence requirement, the normalizing conditions, and the criterion of linguistic invariance were sufficient to justify induction by enumeration as the basic inductive rule for inferring limits of relative frequencies. I no longer hold this view. Alternative inductive rules that are not eliminated by these considerations can be formulated. Ian Hacking has shown, for instance, that rules deviating from induction by enumeration, in a way that depends upon the internal structure of the observed sample, can satisfy all of these requirements.[137] He has not, however, produced a rule that he or anyone else would be anxious to defend. Carnap, however, has been working on serious proposals that deviate from induction by enumeration and cannot be eliminated by the above criteria.[138]

Using the three criteria, I have shown that the "corrective" term c cannot be a function of n alone or of m_i/n alone. The criterion of linguistic invariance rules out any c that is a function of i or k (alone or in combination with other variables) if k is the number of *predicates* and i is an index of *predicates*. Predicates are linguistic entities—names of attributes or properties—so their number and arrangement can be altered by arbitrary linguistic manipulations. If, however, c is taken to be a function of the *attributes* themselves, rather than names of them, the criterion of linguistic invariance is inapplicable. Attributes, in contrast to predicates, are not linguistic entities, so their nature is not affected by linguistic manipulations.

Although the problem of the "corrective" term is not entirely resolved, I find it difficult to conceive of an inductive method, embodying a nonvanishing "corrective" term that is a function of the attributes themselves, that is not perniciously arbitrary in some fashion. In his excellent survey of a continuum of inductive methods, Carnap showed how it is possible to isolate an *empirical factor* and a *logical factor*.[139] The empirical factor is the observed frequency; the logical factor is the "corrective" term. If the logical factor depends upon the language, the criterion of linguistic invariance will eliminate the rule. The logical factor might, however, depend upon the attributes themselves. Suppose, for example, that the logical factor depends upon the number of attributes that are exemplified anywhere in the universe. We could hardly be expected to know that number prior to all applications of inductive logic. Suppose instead that the logical factor depends upon the number of attributes in a "family"; in our examples, the colors constitute a family of attributes. Again, it would seem that any partition of the color spectrum into separate color attributes would involve some sort of arbitrariness. Moreover, even if we know somehow the precise number of color properties, the color spectrum could be partitioned in infinitely many ways to yield the correct number of divisions. If it is not the number of properties we need, but rather some characteristics of these properties, things become no easier. We may think of green as a cool color and red as a hot color, but what possible bearing could these characteristics of the colors have upon the frequency with which they occur? These characteristics of the colors are nonfrequency characteristics—in sharp contrast to the fact that green is a color you run into a lot in nature, while red is comparatively rare. It would seem to follow, from Hume's arguments concerning our inability to know a priori about the connections between distinct properties, that we cannot know anything about the relations between frequency and nonfrequency characteristics of colors prior to the use of induction. What kind of knowledge could it be? The nonfrequency characteristics of properties are to be built into the very inductive rules for inferring frequencies, and these rules are designed to produce useful knowledge of frequencies. Is the "corrective" term an expression of a new kind of synthetic a priori proposition concerning the relation between the phenomenological characteristics of color attributes and the frequencies with which these attributes occur? I am deeply troubled by the logical interpretation of probability, for it seems to escape the pernicious arbitrariness of linguistic variance only by embracing what may turn out to be an even more pernicious apriority.

The personalistic theorist has a much easier answer to the question about the "corrective" term. We need not, he would say, imagine ourselves facing a question about frequencies with a blank mind which is only capable of recording observations and making inductive inferences. We have some sort of prior opinion about the frequencies we are trying to infer. We use observational data to modify that prior opinion. If the observed frequency does not coincide with the prior opinion, the prior opinion is able to supply the "corrective" term. The details of the inference are given by Bayes' theorem, which will be discussed more fully below.

The frequency theorist is by no means stuck with the observed frequency as the inferred value of the limit of the relative frequency in all cases. Induction by enumeration is a method to be applied only where we have no information beyond the observed frequency in the sample upon which to base our inference. Even in that case, it is perfectly consistent to maintain that probability is, by definition, the limit of the relative frequency, but probabilities are to be ascertained by some rule besides induction by enumeration. If, however, induction by enumeration is adopted as the basic inductive rule, it is still subject to correction. Suppose, for instance, that a die has been tossed thirty times and the side one has come up six times for a relative frequency of one fifth. We examine the die and see that it appears to be symmetrical, and the tossing mechanism appears to be unbiased. We do *not* conclude that the limit of the relative frequency of side one is one fifth, for we have a great deal of other experience with the tossing of symmetrical objects, and this experience confirms by and large the view that the alternatives are equiprobable. This is not a regression to the classical interpretation. It is an inductive inference from a large body of frequency information. In such cases the inference from the large body of frequency data supersedes the inference by induction by enumeration from the more restricted data regarding only the class of tosses of that particular die.

Before leaving the discussion of rules for inferring limits of relative frequencies, I must explain one serious argument against induction by enumeration. This point has been made by Carnap. We have already seen that an incoherent betting system is one in which the bettor must lose no matter what happens. The requirement of coherence guards against this situation. Another undesirable type of betting system is one in which the bettor may lose, but he cannot win. This is not quite as bad as incoherence, for it does leave open the possibility that he will come out even. As long, however, as winning is an impossibility, a system of bets is irrational. Carnap therefore sets up a stronger requirement than coherence; it is known as *strict coherence*.[140] Induction by enumeration, it turns out, seems to violate this condition. Suppose that all observed *A* have been *B*, so the observed frequency is one. That is the value we infer for the limit. If this value is translated into betting odds, it means that the bettor will risk *any stake whatever* against an opponent's stake of zero value that the next *A* will be a *B*. This is surely an irrational kind of bet. If the bettor "wins" he wins nothing, whereas if he loses, he loses a stake that has a value.

In view of this undesirable feature of induction by enumeration, we might be tempted to introduce the "corrective" term simply as a safety factor. Its function would be to keep inferred values away from the extremes of zero and one. It is hard to know exactly what form the safety factor should take. If it is too cautious it will make us pass up favorable bets because of the risk; if it is too liberal it will lead us to make rash bets. The fact of the matter is, of course, that a statement about the limit of the relative frequency is synthetic, and we can never be sure of the truth of a synthetic assertion. Hence, we must not take any such inferred value and use it uncritically to determine betting odds. Other factors enter in, especially the amount of inductive evidence we have to support the inference in question.

VII. The Confirmation of Scientific Hypotheses

Quite early in this essay, I acknowledged the fact that induction by enumeration is a far cry from what we usually regard as scientific inference. When we think of scientific reasoning, we are likely to bring to mind the grand theories like those of Galileo, Newton, Darwin, or Einstein, and to contemplate the manner in which they were established. This is in obvious contrast to the attempt to infer the limit of a relative frequency from the observed frequency in an initial section of a sequence of events. Scientific inference is usually thought to be hypothetico-deductive in structure. Induction by enumeration is puerile, as Francis Bacon remarked, and the hypothetico-deductive method is regarded as a great improvement over it. As we saw earlier, however, the hypothetico-deductive method is a mode of *ampliative* inference, and this warrants our treating it as a species of induction. It differs in fundamental respects from deduction.

During the nineteenth century, two ways of contrasting induction and deduction gained some currency, and they are still with us. Both arose from a consideration of the hypothetico-deductive method, and both are fundamentally mistaken. First, induction was held to be the inverse of deduction. In the hypothetico-deductive schema, a *deduction* from premises to conclusion establishes a prediction to be confronted with empirical fact. If the prediction happens to be true, an *induction* from the conclusion to the premises confirms the hypothesis.[141] Second, it was also held that deductive inference is a method of justification, while inductive inference is a process of discovery. The deduction of a prediction from a hypothesis and initial conditions is the heart of the inference by which we test hypotheses. The inductive inference is taken to be the process of trying to think up an appropriate hypothesis to serve as a premise in the foregoing deduction. William Whewell called it "guessing."[142]

The distinction between discovery and justification is extremely important, but it is not coextensive with the distinction between induction and deduction. Our earlier discussion of the hypothetico-deductive method gives ample reason for refusing to merge these two distinctions. The justification of hypotheses is not purely deductive. Even after a hypothesis has been thought up, there is still a nondemonstrative inference involved in confirming it. To maintain that the truth of a deduced prediction supports a hypothesis is straightforwardly inductive. Leaving the problem of discovery entirely aside, we must still separate deductive and inductive elements of scientific inference.

The view that induction is the inverse of deduction appears to be based upon an extremely widespread misconception concerning the relations between induction and deduction. This notion may not often be explicitly stated and defended, but it seems to arise easily if one reflects a little upon the nature of logic. It may rank as the leading unconscious misconception regarding induction. This view takes inductions to be defective deductions—deductions that do not quite make the grade. An induction, according to this notion, is some sort of approximation that does not fully achieve the status of valid deduction. Inductive inferences are seen as fallacies we are not quite willing to reject outright; they are more to be pitied than condemned.

Pervasive misconceptions usually have an element of truth, and this one is no exception. Deductions *are* limiting cases of inductions in certain respects. The logical necessity relating premises and conclusion in valid deduction can be regarded as the limiting case of the high probabilities we attempt to achieve for our inductions.[143] At the same time, the emptiness of valid deduction is also a limiting case of decreasing the ampliative character of inductions, but this side of the coin seems not to be as frequently noticed. The main trouble with the "almost-deduction" theory of induction is that it does not furnish a concept of approximation to deduction that enables us to distinguish good inductions from plain logical errors. If anything, it tends to direct attention away from finding one. Instead of motivating a careful logical analysis of induction, it tends to make us think we should behave like social workers, providing underprivileged inductive inferences with the necessities enjoyed by valid deductions.

Let me take a moment to mention some of the defects from which these inferences that do not quite make the grade may suffer. For one thing, an inference may be an *enthymeme*—i.e., a valid deduction with a suppressed premise. Inductions have this characteristic; they can be transformed into valid deductions by supplying a suitable premise. The most monstrous *non sequitur* ever to find its way into a freshman theme can also be transformed into a valid deduction in this way! Enthymematic character certainly cannot serve as a criterion of inductive correctness.

Another defect consists in having premises that differ only slightly from those required to satisfy a valid deductive schema. For instance, an almost universal premise may be all we have when a strictly universal premise is needed. Thus, we may think that contraposition, although not strictly valid, is inductively sound. We may think it a good induction to infer "Almost all philosophers are unkind" from "Almost all kind people are nonphilosophers." Unfortunately, the moment we depart from strict universality in our premises we forego any semblance of logical correctness, deductive or inductive. In this case it is an all-or-none affair; approximation does not help.[144]

Most deductive fallacies that have been named and catalogued bear some resemblance to valid deductive forms. Furthermore, they are arguments people are tempted upon occasion to accept. When one becomes aware that his pet argument commits a common fallacy—say, the fallacy of the undistributed middle—the obvious move is to claim that it was never meant as a valid deduction, but only as an induction. In this way, deductive fallacies become, *ipso facto,* correct inductions. "All logicians are mongolian idiots" thus qualifies as a sound inductive conclusion from the premises "All logicians are living organisms" and "All mongolian idiots are living organisms."

The hypothetico-deductive method is another type of argument that seems to approximate deductive validity. From a hypothesis, in conjunction with statements of initial conditions whose truth is not presently being questioned, a prediction is deduced. Observation reveals that the prediction is true. We conclude that the hypothesis is confirmed by this outcome. The inference is, as certain nineteenth-century theorists insisted, an inverse of deduction. By interchanging the conclusion with one of the premises it can be transformed into a valid deduction. Without the interchange, however, the inference goes from conclusion to premise, for we seek to establish the hypothesis on the ground of the true prediction. It is a deductive fallacy closely akin to affirming the consequent. These are not adequate credentials for admission into the class of correct inductions.

Questions of deductive validity are generally referred to systems of formal logic, and they usually admit definite and precise answers. Questions of inductive correctness are far more frequently answered on an intuitive or common-sense basis. Although there have been various efforts, of which Carnap's is the leading example, to formalize inductive logic, such systems have not gained wide acceptance. In spite of their existence, few important questions are decided on a formal basis. Actually, however, the mathematical calculus of probability itself is an invaluable formal tool which is too often ignored in the treatment of problems in inductive logic.[145] It provides, as I shall try to show, important insight into the structure of scientific inference. In order to prepare the ground for the application of the mathematical calculus of probability to the problem of the confirmation of scientific hypotheses, I shall first discuss two important contemporary attacks upon the hypothetico-deductive method.

1. Hanson's Logic of Discovery. Even when we recognize that inductive inference is not properly characterized as a process of discovery, and even if we admit the existence of an inductive logic of justification, there still remains the important question of whether any kind of logic of discovery can exist. The received answer among contemporary philosophers of science is negative. The process by which we think up scientific hypotheses is, they say, strictly a psychological affair. It is not and cannot be reduced to a set of rules of procedure. The discovery of hypotheses requires insight, ingenuity, and originality. The process of finding answers to scientific questions cannot be transformed into a mechanical routine. Science, they say, is not a sausage machine into which you feed the data and by turning a crank produce finished hypotheses.

The standard answer is, nevertheless, a very disappointing one. It is frustrating for someone who is seriously grappling with a difficult scientific problem to be told that logic has no help whatever to offer him in thinking it through. Only after the interesting, original, creative work is done can logic be brought to bear. According to the received opinion, logical analysis can be used for the dissection of scientific corpses, but it cannot have a role in living, growing science. This view relegates philosophy of science to an intolerable position in the eyes of some philosophers. In protest, N. R. Hanson has tried to show that logic has bearing upon the unfinished business of science by arguing that, in addition to the admittedly psychological aspects of scientific innovation, certain logical considerations properly enter into the discovery of hypotheses.[146]

The issue can be thrown into relief by considering the situation in deductive logic, where important answers have been rigorously established.[147] The problem is familiar from elementary mathematics. Some kinds of problems can be solved by following a routine procedure; others cannot. If such a procedure does exist, we say that there is an *algorithm.* The most obvious example is that differentiation is routine; integration is not. In deductive logic we can distinguish several situations. First, given a set of premises, we may be asked to find a valid conclusion of some particular sort. For example, given two categorical propositions, there is a mechanical method for finding all valid syllogistic conclusions. Such methods exist only for very restricted realms of logic. Second, given a set of premises and a conclusion, we may be asked to determine whether that conclusion follows validly from those premises. In this case, we are not asked to discover the conclusion, but we are, in effect, being asked to discover a proof. If there is a mechanical method for answering this kind of question, we say that a *decision method* exists. For the propositional calculus we have a decision method in the truth tables, but there is no decision method for the whole of the lower functional calculus. It is not merely that we have failed to devise a decision method; the impossibility of a decision method has been proved.[148] Third, given a set of premises, a conclusion, and an alleged demonstration of the conclusion, we may be asked to determine whether the demonstration is valid. This is the kind of question deductive logic is designed to answer. Such a logic is a logic of justification, not a logic of discovery. The question of the existence of a deductive logic of discovery can be stated precisely and answered unambiguously.

Turning to the problem of a logic of discovery for empirical science, we must be careful not to pose the question in an unreasonable way. To suggest that there might be a mechanical method that will necessarily generate true explanatory hypotheses is a fantastic rationalistic dream. Problems of discovery completely aside, there is no way of determining for certain that we have a true hypothesis. To make such a demand upon a logic of discovery is obviously excessive. Not since Francis Bacon has any empiricist regarded the logic of science as an algorithm that would yield all scientific truth.

What might we reasonably demand of our logic of discovery if there is to be such a thing? Hanson, and Peirce before him, answer not that it must generate true hypotheses, but that it should generate *plausible conjectures.* Hanson believes this demand can be fulfilled. He begins by distinguishing "reasons for accepting a hypothesis *H*" from "reasons for suggesting *H* in the first place." He elaborates as follows:

> What would be our reasons for accepting *H*? These will be those we might have for thinking *H* true. But the reasons for suggesting *H* originally, or for formulating *H* in one way rather than another, may not be those one requires before thinking *H* true. They are, rather, those reasons which make *H* a *plausible type of conjecture.*[149]

Philosophers who have argued against the existence of a logic of discovery have maintained that the process of discovery is governed entirely by psychological factors. Hanson readily admits the existence of nonlogical aspects of the discovery of hypotheses, but he claims that there are, in addition, perfectly good logical reasons for regarding hypotheses of a particular type as those most likely to succeed. These reasons are logically distinct from the kinds of reasons that later, in the case of successful hypotheses, make us elevate the hypothesis from the status of plausible conjecture to the status of acceptable, true, or highly confirmed hypothesis. Hanson continues:

> The issue is whether, *before* having hit a hypothesis which succeeds in its predictions, one can have good reasons for anticipating that the hypothesis will be one of some particular *kind.* Could Kepler, for example, have had good reasons, before his elliptical-orbit hypothesis was established, for supposing that the successful hypothesis concerning Mars' orbit would be of the noncircular kind?[150]

There is a crucial switch in these two sentences. In the first, Hanson refers to what happens before we have "hit" a hypothesis; this means, I take it, before it came to mind. In the second sentence, he discusses what happens before a hypothesis is "established," but not necessarily before anyone thought of it. There is, presumably, a time between first thinking of a hypothesis and finally accepting it during which we may consider whether it is even plausible. At this stage we are trying to determine whether the hypothesis deserves to be seriously entertained and tested or whether it should be cast aside without further ceremony.

I do not want to be misunderstood as attempting a historical or psychological account of the actual course of scientific thought. The point is strictly logical. There are, it seems to me, three logically distinct aspects of the treatment of scientific hypotheses. It is easy to talk in terms of a temporal sequence of steps, but this is merely a manner of speaking. It does not matter which comes first or whether they are, in fact, mixed together. There are still three distinct matters: (1) thinking of the hypothesis, (2) plausibility considerations, and (3) testing or confirmation.

Hanson has argued (correctly I think) that there is an important logical distinction between plausibility arguments and the testing of hypotheses, but he has (mistakenly I think) conflated plausibility arguments with discovery. Continuing with Kepler as an example, Hanson discusses hypotheses that would have been rejected by Kepler as implausible.

> Other *kinds* of hypotheses were available to Kepler: for example, that Mars' *color* is responsible for its high velocities, or that the dispositions of Jupiter's moons are responsible. But these would not have struck Kepler as capable of explaining such surprising phenomena. Indeed, he would have thought it *un*reasonable to develop such hypotheses at all, and would have argued thus.[151]

Kepler would, quite plainly, have rejected such hypotheses *if they had occurred to him.* There is no reason to suppose, however, that these considerations were psychologically efficacious in preventing Kepler from thinking of such hypotheses (although they might have been efficacious in preventing him from mentioning them) and in causing him to think of others instead. Furthermore, it does not matter in the slightest. What does matter is that, had such unreasonable hypotheses crossed Kepler's mind, plausibility arguments would have sufficed to prevent them from coming to serious empirical testing.

One basic question remains. Plausibility arguments have been distinguished from hypothesis testing and confirmation on the one hand and from the psychology of discovery on the other. What, precisely, is their status? Are plausibility considerations psychological or subjective in character? Do they play a legitimate role in science, or do they merely express the prejudices of the scientist or the scientific community? Are they different in kind from the considerations involved in the confirmation of hypotheses? An answer to this question will be forthcoming when we look more closely at what the probability calculus tells us about confirmation.

2. *Popper's Method of Corroboration.* In my earlier discussion of Karl Popper's attempt to avoid the problem of induction, I explained his

rejection of induction by enumeration and his denial that the hypothetico-deductive method is a suitable way of confirming scientific hypotheses. Popper maintains it is *not* the function of science to produce highly probable hypotheses or hypotheses that are highly confirmed by the evidence. If it were, the way would be left open for relatively vacuous hypotheses that are better classed as metaphysics than as science. The aim of science is rather to find hypotheses that have great content and make important assertions about the world. Such hypotheses are bold conjectures, and their very boldness makes them highly falsifiable. Moreover, every effort must be made to find evidence that does falsify such hypotheses. A highly falsifiable hypothesis that has withstood serious efforts at falsification is highly corroborated.

The process of corroboration bears some resemblance to confirmation. An unsuccessful attempt to falsify a hypothesis is precisely what the hypothetico-deductive theorist would identify as a positive confirming instance. There is, however, a crucial difference. The hypothetico-deductive theorist attempts to start with probable hypotheses and find further support for them through positive confirmations. If more than one hypothesis is available to explain all the available data, the hypothetico-deductivist would choose the most probable one. Popper's method of corroboration, by contrast, tries to begin with the least probable hypothesis, for probability is related inversely to content.[152] It seeks to falsify this hypothesis. Failure to do so tends to increase the degree of corroboration. If more than one hypothesis remains unfalsified, we select the least probable one. Thus, one might say, Hanson attacks the hypothetico-deductive method for failure to take account of *plausibility* arguments, while Popper attacks the same method for failure to incorporate *implausibility* considerations. While it appears that Hanson's attack and Popper's attack are mutually incompatible, I shall try to show that each one has a valid foundation, and each points to a fundamental shortcoming of the hypothetico-deductive approach. I shall argue that the logical gaps in the hypothetico-deductive method can be filled by means of the ideas suggested by Hanson and Popper and that these ideas lead us to indispensable, but often neglected, aspects of the logic of scientific inference.

3. *Bayesian Inference.* The basic trouble with the hypothetico-deductive inference is that it always leaves us with an embarrassing superabundance of hypotheses. All of these hypotheses are equally adequate to the available data from the standpoint of the pure hypothetico-deductive framework. Each is confirmed in precisely the same manner by the same

evidence.[153] An hypothesis is confirmed when, in conjunction with true statements of initial conditions, it entails a true prediction. Any other hypothesis that, in conjunction with (the same or different) true statements of initial conditions, entails the same prediction is confirmed in the same way by the same evidence. It is always possible to construct an unlimited supply of hypotheses to fill the bill. It is essentially a matter of completing an enthymeme by supplying a missing premise, and this can always be done in a variety of ways. The hypothetico-deductive method is, therefore, hopelessly inconclusive for determining the acceptability of scientific hypotheses on the basis of empirical data. Something must be done to improve upon it.

It is at this point that the probability calculus can come to our aid. In a preceding section I showed how a simple form of Bayes' theorem follows from axioms for the probability calculus: [154] If $P(A, C) \neq 0$,

$$P(A \cap C, B) = \frac{P(A, B) \times P(A \cap B, C)}{P(A, C)}$$
$$= \frac{P(A, B) \times P(A \cap B, C)}{P(A, B) \times P(A \cap B, C) + P(A, \bar{B}) \times P(A \cap \bar{B}, C)}.$$

As a theorem in the uninterpreted calculus of probability, it is entirely noncontroversial. It was concretely illustrated by means of a simple example. Its application to examples of that type is also straightforward. Now, let me stretch and bend the meanings of words a bit in order to begin an explanation of the application of Bayes' theorem to the problem of the confirmation of scientific hypotheses. In the previous example, we might say, Bayes' theorem was used to calculate the probability of a "cause" and to assign a probability to a "causal hypothesis." A red card was drawn, and we asked how this "effect" came about. There are two "possible causes"—throwing a one and drawing from the half red deck or throwing some other number and drawing from the largely black deck. While we are still abusing causal language, let us describe the probabilities required to calculate the probability of the "causal hypothesis." We need two other probabilities in addition to $P(A \cap B, C)$, the probability of the "effect" given the "causal hypothesis." We need $P(A, B)$, the *prior probability* of the "cause," and we need $P(A, C)$ or $P(A \cap \bar{B}, C)$, depending on the version of Bayes' theorem we pick. $P(A, C)$ is the probability of the "effect" regardless of the "cause" from which it issues; $P(A \cap \bar{B}, C)$ is the probability of the "effect" if our "causal hypothesis" is false.

There is no difficulty in understanding all these probabilities in our simple game of chance, but things get much more complex when we try to apply Bayes' theorem to genuine scientific hypotheses. Serious problems of interpretation arise. I shall claim, nevertheless, that Bayes' theorem provides the appropriate logical schema to characterize inferences designed to establish scientific hypotheses. The hypothetico-deductive method is, I think, an oversimplification of Bayes' theorem. It is fallacious as it stands, but it can be rectified by supplementing it with the remaining elements required for application of Bayes' theorem.

Let us, therefore, compare the hypothetico-deductive method with Bayes' theorem. From an hypothesis H and statements of initial conditions I, an observational prediction O is deducible. For purposes of this discussion we assume I to be true and unproblematic. Under this assumption H implies O. We can provide a loose and preliminary interpretation of Bayes' theorem, even though many difficult problems of interpretation remain to be discussed. Let "A" refer to hypotheses like H; let "B" refer to the property of truth; and let "C" refer to the observed result with respect to the prediction O. If positive confirmation occurs "C" means that O obtains; in the negative case "C" designates the falsity of O. This interpretation makes the expression on the left-hand side of Bayes' theorem refer to precisely the sort of probability that interests us; "$P(A \cap C, B)$" designates the probability that a hypothesis of the sort in question, for which we have found the given observational result, is true. This is the probability we are looking for when we deal with the confirmation of scientific hypotheses.

In order to compute the value of $P(A \cap C, B)$, the *posterior probability* of our hypothesis, we need, as we have seen, three probabilities. The hypothetico-deductive method provides only one of them. Given that H implies O and that O obtains, $P(A \cap B, C) = 1$. Inspection of Bayes' theorem reveals, however, that this value is entirely compatible with a small posterior probability for the hypothesis. A small value for $P(A, B)$ and a large value for $P(A \cap \bar{B}, C)$ nullify any tendency of the confirmation to enhance the value of $P(A \cap C, B)$. Successful confirmation requires attention to all three required probabilities, only one of which is provided by the hypothetico-deductive argument. Notice, however, that Bayes' theorem embodies the asymmetry between confirmation and falsification. If H implies O and O does not obtain, then $P(A \cap B, C) = 0$, and it follows immediately that the posterior probability of the hypothesis, $P(A \cap C, B)$, likewise equals zero. Falsification holds a special place in the logic of scientific inference, as Popper has emphasized.[155]

We are left with the task of interpreting the remaining probability expressions so that they will have meaning for the logic of scientific inference. Consider the prior probability $P(A, B)$. It is the probability that our hypothesis is true regardless of the outcome of our prediction. This probability is logically prior to the empirical test provided by the hypothetico-deductive method. How are we to make sense of such a probability? Regardless of our detailed answer, one preliminary point is apparent. Prior probabilities fit the description of Hanson's plausibility arguments. Plausibility arguments embody considerations relevant to the evaluation of prior probabilities. They are logically prior to the confirmatory data emerging from the hypothetico-deductive schema, and they involve direct consideration of whether the hypothesis is of a type likely to be successful. These plausibility arguments do not, of course, constitute a logic of discovery. *They are not only admissible into the logic of justification; they are an indispensable part of it.* Bayes' theorem requires the prior probabilities as well as the confirmatory data, so plausibility arguments as well as hypothetico-deductive arguments are essential elements of a logic of scientific inference. We shall have to discuss these plausibility arguments with more precision, but we have at least succeeded in locating them in the general schema.

The denominator of Bayes' theorem can be written in either of two ways because of the theorem on total probability.[156] The simpler form involves $P(A, C)$, the probability of obtaining the observational result regardless of the truth of our hypothesis H. The more complex form requires $P(A, \bar{B})$ and $P(A \cap \bar{B}, C)$. $P(A, \bar{B})$ is logically linked with $P(A, B)$, so it involves nothing new. $P(A \cap \bar{B}, C)$ is logically independent of $P(A \cap B, C)$; it is the probability of getting our observational result if the hypothesis H is false. Either form of the theorem makes it obvious that we must consider the probability that our prediction would come true even if our hypothesis were false. Other things being equal, the less probable our observational result if the hypothesis is false, the more this observational result confirms the hypothesis.

I have already discussed Popper's eloquent plea for the view that scientific hypotheses, to be worthwhile, must run the risk of falsification. The more falsifiable they are, and the more strenuously we have tried to falsify them, the better they are, as long as they survive the tests without being falsified. Popper maintains that the more falsifiable they are the less probable they are (and this is a *prior* probability). To the extent that hypothetico-deductive theorists have been aware of prior probabilities, they have claimed that hypotheses are better confirmed if they have higher prior probabilities—i.e., if they are more plausible. Popper claims better corroboration for hypotheses that are more audacious and less plausible.

I cannot accept Popper's view that we ought not to be concerned with confirming hypotheses and enhancing their posterior probabilities. It seems to me that Bayes' theorem gives us an unequivocal answer to the question of whether we ought to regard high prior probability as an asset or a liability. Plausibility contributes positively to the acceptability of hypotheses. Nevertheless, Popper has, it seems to me, a fundamental insight. There is another way for a hypothesis to run a risk of falsification, and this is revealed by Bayes' theorem. A hypothesis risks falsification by yielding a prediction that is very improbable unless that hypothesis is true. It makes a daring prediction, for it is not likely to come out right unless we have hit upon the correct hypothesis. Confirming instances are not likely to be forthcoming by sheer chance. This state of affairs is reflected in a small value for $P(A \cap \bar{B}, C)$. The hypothesis that runs this kind of risk of falsification without being falsified gains more in posterior probability than one that runs less of such risk. This does not mean, however, that the hypothesis itself must be implausible. A small value for $P(A \cap \bar{B}, C)$ is perfectly compatible with a large value for $P(A, B)$. This question of falsifiability is nicely illustrated by an example from the history of optics. Early in the nineteenth century, when the wave theory of light was coming into its own, Poisson deduced as a consequence of that theory that the shadow of a disc should have a bright spot in its center. Poisson regarded this derivation as a *reductio ad absurdum* of the wave theory, but Arago was later able to announce triumphantly that a positive result was obtained when the experiment had been performed. The wave theory had been confirmed! [157] This was indeed an impressive confirmation, for the predicted consequence surely seemed utterly unlikely on any other hypothesis. It is not that the wave theory itself was so improbable; the thing that was really improbable was the occurrence of the bright spot in the middle of the shadow *if the wave theory were not true.*

Compare the foregoing example with one of the opposite sort. About fifteen years ago a pseudo-psychological theory known as *dianetics* gained considerable popularity.[158] This theory embodied an explanation of psychological disorders and recommended a course of treatment. Dianetic therapy was widely practiced, and it seems undeniable that a number of "cures" were effected. People with neurotic symptoms who underwent the prescribed treatment exhibited definite improvement.

These results must be considered confirming evidence for dianetic theory, but they lend very little support to it. The trouble is that the same results are very probable even if the dianetic hypothesis is false, so $P(A \cap \bar{B}, C)$ is high. It is well known that many psychological disorders are amenable to faith healing. They can be cured by *any* method the patient sincerely believes to be effective. There is no doubt that many people had great faith in dianetics, so faith healing constitutes a better explanation of the cures than does the dianetic "hypothesis" itself.

In most of Popper's statements about probability and content he makes it fairly clear that he regards a low *prior* probability as a desirable feature in a scientific hypothesis.[159] There is one passage, however, in which he seems strongly to suggest that he is referring, not to prior probability, but to the probability $P(A \cap \bar{B}, C)$ of the experimental result in case the hypothesis is false. This is the probability for which, in contrast to the prior probability of the hypothesis, a *low* value tends to enhance the posterior probability of the hypothesis.

> A theory is tested not merely by applying it, or by trying it out, but by applying it to very special cases—cases for which it yields results different from those we should have expected without that theory, or in the light of other theories. In other words we try to select for our tests those crucial cases in which we should expect the theory to fail if it is not true. Such cases are "crucial" in Bacon's sense; they indicate the crossroads between *two* (or more) theories. For to say that without the theory in question we should have expected a different result implies that our expectation was the result of some other (perhaps an older) theory, however dimly we may have been aware of this fact. But while Bacon believed that a crucial experiment may establish or verify a theory, we shall have to say that it can at most refute or falsify a theory. It is an attempt to refute it; and if it does not succeed in refuting the theory in question—if, rather, the theory is successful with its unexpected prediction—then we say that it is corroborated by the experiment. (It is the better corroborated the less expected, or the less probable, the result of the experiment has been.)[160]

I have quoted this passage, not to try to reveal any inconsistency in Popper's writings, but rather to show how admirably his conception, as expressed in the foregoing remarks, fits the Bayesian schema. The quoted statement does a splendid job of describing the concrete example from optics.

Bayes' theorem casts considerable light upon the logic of scientific inference. It provides a coherent schema in terms of which we can understand the roles of confirmation, falsification, corroboration, and plausibility. It yields a theory of scientific inference that unifies such apparently irreconcilable views as the standard hypothetico-deductive theory, Popper's deductivism, and Hanson's logic of discovery. However, it still poses enormous difficulties of interpretation. We have been concerned so far mainly with the formal characteristics of Bayes' theorem, and the hints at interpretation have been purposely vague. The formal schema requires prior probabilities, but what precisely are they? To link them with Hanson's plausibility arguments does not get us far. The notion of prior probability cries out for further clarification, and it must be sought in the light of the interpretation of the probability concept in general. I shall discuss this issue from the standpoint of each of the three leading interpretations presented above.

1. According to the logical interpretation, as we have seen, probability is fundamentally an a priori measure of possible states of affairs. The state descriptions provide a list of all possible states of the universe, and weights are assigned to them. A scientific hypothesis will be true if certain of these state descriptions hold, but false if others do. The set of all state descriptions compatible with the hypothesis is its range, and the prior probability of the hypothesis is the sum of the values attached to the state descriptions in its range. The accumulation of observational evidence enables us to calculate posterior probabilities of hypotheses in accordance with Bayes' theorem, and the prior probabilities are available as required. For the reasons already stated, I reject this interpretation precisely because it embodies a priori prior probabilities. They play an indispensable role in determining the probabilities of factual hypotheses, and their status is extremely dubious.

2. According to the personalistic interpretation, probabilities are simply degrees of belief in the truth of statements. The probability calculus imposes conditions upon the relationships among these various degrees of conviction, but if they do not violate these conditions they are rational. That is all the personalistic interpretation requires. Prior probabilities are totally unproblematic for the personalist. When a hypothesis is entertained we have a certain degree of conviction in its truth. It does not matter whether this is based upon solid evidence, sheer prejudice, or unfettered emotion. This *is* the prior probability of the hypothesis. Further experience may affect this degree of belief, thus issuing in posterior probabilities. Bayes' theorem expresses the relations that must hold among these various degrees of belief if the probability calculus is not to be violated. Theorists who cannot swallow a priori prior probabilities may find it difficult to see where prior probabilities of any other kind are to be found. The personalistic theorists answer this question for them. Prior opinion is always available, so the prior probabilities required by

Bayes' theorem are never lacking. So completely are the personalists wed to Bayes' theorem that they have even taken its name and are now known as "Bayesians." [161]

An examination of Bayes' theorem reveals the fact that a prior probability of zero or one determines by itself the same value for the posterior probability. In the remaining cases, the prior probability has only a part in determining the posterior probability. Under some rather mild assumptions, the role played by the prior probabilities becomes smaller and smaller as observational evidence increases. This fact has been rightly accorded a central place in the arguments of the Bayesians. We come to any problem, according to the personalistic theorist, with opinions and preconceptions. The prior convictions of reasonable people can differ considerably. As these individuals accumulate a shared body of observational evidence, the differences of opinion will tend to disappear and a consensus will emerge. The influence of the prior opinion will fade in the face of increasing evidence if the prior opinions do not have the extreme values zero and one. It is not necessary that these individuals be genuinely open-minded about the various hypotheses; it is enough if their minds are slightly ajar.[162] Before the advent of the personalistic theory, there was great reluctance to admit that Bayes' theorem could be applied at all in dealing with the confirmation of scientific hypotheses. The trouble seemed to lie with the prior probabilities. Any way of handling them seemed to make them unacceptably a priori or subjectively slippery. Methods of confirmation which would not require these unrespectable entities were sought. The Bayesian attitude toward this program is nicely captured in a paraphrase of a statement by de Finetti, the foremost contemporary personalistic theorist: "People noticing difficulties in applying Bayes' theorem remarked, 'We see that it is not secure to build on sand. Take away the sand, we shall build on the void.' " [163] The personalist, however, rejects the methods that ignore prior probabilities, and he willingly embraces them with their full subjectivity. As an excellent recent account of the Bayesian approach puts it: "Reflection shows that any policy that pretends to ignore prior opinion will be acceptable only insofar as it is actually justified by prior opinion." [164] By showing how the use of Bayes' theorem leads to substantial intersubjective agreement, the personalists argue that the subjectivity of prior probabilities is not pernicious.

I enthusiastically applaud the emphasis the personalistic theorists have placed upon the use of Bayes' theorem, but I cannot accept their far-reaching subjectivism. Although satisfaction of the relations established by the probability calculus is a necessary condition for rationality, it is not a sufficient condition. Other requirements for rational belief need to be found. The Bayesians themselves seem to acknowledge this need when they impose further conditions upon the prior probabilities in order to insure convergence of opinion in the light of evidence. Prior probabilities are not, of course, alone in being subjective. All of the other probabilities that enter into Bayes' theorem are likewise subjective. This includes the probability $P(A \cap B, C)$ that the observational evidence would occur if the hypothesis in question were true, and the probability $P(A \cap \bar{B}, C)$ that it would occur if the hypothesis were false. All these subjective probabilities may actually be based upon extensive observation and inductive generalization therefrom, but they may also be lacking any foundation whatever in objective fact. As nearly as I have been able to tell, there is no reason within the personalistic framework to reject as irrational a set of opinions which conflicts with the bulk of experience and dismisses this fact on the ground that most observation is hallucinatory. Moreover, I cannot see any ground for characterizing as irrational opinions that have arisen out of observation by application of some perverse inductive method. Although the personalist can reject a series of opinions based upon the counterinductive rule because it violates the normalizing conditions, it is easy to formulate a normalized counterinductive rule that satisfies these conditions but still makes past experience a negative guide to the future:

$$\text{From } F^n(A, B) = \frac{m}{n}, \text{ to infer } \lim_{n \to \infty} F^n(A, B) = \frac{1}{k-1} \times \frac{n-m}{n}$$

The fact that this rule is not asymptotic would not invalidate its use on strictly personalistic grounds. Personalistic theorists do not actually condone misuses of experience in either of the foregoing ways, but the principles by which they avoid them need to be spelled out, examined, and justified.

3. As we have seen, the frequency interpretation defines probability as the limit of the relative frequency with which an attribute occurs in an infinite sequence of events. This definition gives rise immediately to the problem of application of probability to the single case. In my earlier discussion of the frequency theory I outlined a way of dealing with this problem. At the same time, I mentioned another problem of application that presents difficulties for the frequentist. This is the problem of explicating the logic of the confirmation of scientific hypotheses in terms of frequencies. It has been persuasively argued that the relation between

evidence and scientific hypothesis cannot be understood exclusively in terms of frequency concepts—in particular, it has seemed outrageous to maintain that prior probabilities of scientific hypotheses could be construed as relative frequencies.[165] The two problems are not unrelated, for the probability of hypotheses is an instance of the problem of the single case. Any given scientific hypothesis, as a single entity, is either true or false—just as a single toss of a coin results in a head or does not.

The specification of the attribute class is of no particular difficulty for the frequentist attempting to apply probability to the single case; the whole difficulty rests with the selection of the reference class. The rule is to select the broadest homogeneous reference class available. In the effort to show how the frequency concept of probability can be made relevant to the probability of hypotheses through the use of Bayes' theorem, we must find the appropriate prior probability $P(A, B)$.[166] This is the probability that hypotheses of a certain type are true. The attribute of truth is given directly by the fact that we are looking for true hypotheses. In attempting to choose an appropriate reference class, we are trying to find out what type of hypothesis is likely to be true. This is how Hanson describes the plausibility considerations whose importance he so staunchly defends. A hypothesis that belongs to the class of plausible conjectures is one that has a high prior probability. One that belongs to the class of preposterous conjectures is one that has a vanishingly small prior probability. The interesting question is how we are to determine what considerations are relevant to plausibility or prior probability. Characteristics that are statistically relevant to the truth or falsity of scientific hypotheses are properties that determine a homogeneous reference class. To evaluate a given hypothesis H, we try to find a (practically or epistemically) homogeneous reference class A to which H belongs. A must be a class of hypotheses within which we can say something about the relative frequency of truth. The probability $P(A, B)$ is the probability of truth for hypotheses of this class, and this probability is assigned as a weight to the hypothesis H. This weight, which might be distinguished as a *prior weight*, expresses the plausibility of H.

The characteristics by means of which the reference class A is determined are properties that are logically independent of the confirmatory evidence for the hypothesis H. A prior probability is logically, although not necessarily temporally, prior to the observational verification of the prediction made on the basis of the hypothetico-deductive schema. In many of the interesting cases, the prior probability is not used to determine prior weight; instead, the probability itself is fed into Bayes' theorem along with other probabilities, in order to calculate the posterior probability $P(A \cap C, B)$. This probability yields a *posterior weight* which is based upon plausibility considerations *and* confirmatory evidence.

As examples of the kinds of considerations that serve as a basis for plausibility judgments, Hanson mentions analogy and symmetry. I should like to attempt a larger and more systematic classification. There are, it seems to me, three important types of characteristics that may be used as a basis for plausibility judgments. These characteristics determine the relevant reference class, but they may also be regarded as criteria of plausibility that hypotheses must confront. Success in meeting a given criterion will classify a hypothesis with other plausible hypotheses; failure will group it with implausible ones.

1. Formal criteria. Scientific hypotheses are proposed, not in an epistemic vacuum, but against the background of many previously accepted hypotheses and many that have already been rejected. The newly proposed hypothesis may bear to accepted hypotheses deductive relations that are germane to the plausibility of the new one. If an old hypothesis H_1 entails a new hypothesis H_2, then the *prior probability* of H_2 is at least as great as the *posterior probability* of H_1. If a new hypothesis H_3 is incompatible with an old hypothesis H_4, then the prior probability of H_3 is no greater than the probability that H_4 is false—i.e., one minus the posterior probability of H_4. This point is well illustrated by Velikovski's notorious book, *Worlds in Collision.* This so-called theory, designed to explain certain alleged events as related in the *Old Testament,* entails the falsity of virtually all of modern physics. When the editors of *Harpers Magazine* complained that the scientific world was falling down on its objectivity by refusing to subject Velikovski's "theory" to extensive physical tests, they were overlooking the power and legitimacy of plausibility arguments and prior probabilities.[167]

2. Pragmatic criteria. Although we have all been warned repeatedly about the dangers of confusing the source of a theory with its truth, or the origin with the justification, there are cases in which it is possible to establish a probability relation between the truth of a hypothesis and the circumstances of its discovery. If a religious fanatic without any training in physics or mathematics shows up on our doorstep with a new hypothesis to replace Einsteinian relativity, complaining that most scientists refuse him a fair hearing, we justly place a low estimate on the chances that his hypothesis is true. Considerations of this kind are legitimate only if there is a known probability relation between the character of the individual presenting the hypothesis and the truth of the

hypothesis he advances. If, however, the rejection of the hypothesis is based only upon an emotional reaction to its origin—e.g., the fellow has a beard, which makes a very unfavorable impression—it is flagrantly fallacious.

Pragmatic criteria are, perhaps, less reliable than others, but they are used by scientists, and there is no need to be embarrassed by the fact. They are as objective as any other kinds of considerations. Martin Gardner provides many fascinating examples of the application of pragmatic criteria.[168]

3. Material criteria. Just as relations of entailment or incompatibility can exist between different hypotheses, so too, I think, can there be significant inductive relations among them. This, I suspect, is what Hanson regards as analogy. Certain types of hypotheses have been successful; we may legitimately expect new hypotheses that are similar in relevant respects to be successful as well. Hypotheses are considered plausible, then, on the basis of their analogy with other successful hypotheses. The material criteria encompass those respects in which hypotheses may be relevantly similar to one another.

It is beyond the scope of this essay to attempt any exhaustive list of considerations relevant to the prior probabilities of scientific hypotheses. Relevance is an empirical matter, so the determination of relevant characteristics of hypotheses is a task for empirical science rather than philosophy. It is possible, nevertheless, to indicate what is involved in the material criteria by citing a few familiar examples.

Although no one can say just what simplicity is, everyone seems to agree that it is a very desirable characteristic of scientific hypotheses. A certain type of simplicity lends beauty and elegance to the hypothesis and ease to its application, but such economic and aesthetic considerations are secondary when an issue of truth or falsity is at stake. At this point a different type of simplicity is invoked. We place more confidence in simple than in complex hypotheses for purposes of explaining a given body of fact.[169] We judge the simpler hypothesis more likely to be true. We have learned by experience that this works, and at the same time we have learned by experience to avoid oversimplification.

Hypotheses can sometimes be distinguished through the kinds of causal processes they countenance. For instance, the fight to purge science of its teleological elements has been long and arduous. The explanation of natural phenomena in terms of conscious purposes must have seemed extremely plausible to primitive man. One notable way in which the physics of Galileo and Newton improved upon the physics of

Aristotle was by eliminating the teleological elements in the latter. The admirable success of nonpurposeful explanation in physics provided an important precedent for nonteleological evolutionary theories in biology. The success of these theories in biology has provided a strong basis for assigning low prior probabilities to teleological hypotheses in psychology and sociology. A teleological biological theory like Lecomte du Noüy's *Human Destiny* suffers from a high degree of implausibility.

As a final example, let me sketch some plausibility arguments relating to the nature of physical space. Nothing could seem less plausible to primitive man than the idea that space is homogeneous and isotropic. Everyday experience incessantly confirmed the notion that there is a physically preferred direction. This theory was charmingly elaborated by Lucretius, who held that the primordial state of the universe consisted of all the atoms falling downward. Moreover, this theory implied that space is inhomogeneous. A distinction among different locations was required to support the theory of uniform downward motion as opposed to rest. The doctrine of inhomogeneity and anisotropy persisted in the geocentric cosmologies, and in the early heliocentric ones as well. By Newton's time it had waned considerably. On the supposition that space is Euclidean and lacking in privileged locations or directions, a rather strong plausibility argument can be made for the inverse square law of gravitational attraction. Since the surface of a sphere is proportional to the square of its radius, the supposition that the gravitational force "spreads out" uniformly through space leads to the conclusion that it is inversely proportional to the square of the distance.

These considerations of homogeneity and isotropy extend to the twentieth century and underlie the plausibility arguments for special relativity. As Adolf Grünbaum has convincingly argued, Einstein saw in his famous principle of relativity a notion so plausible that it demanded formulation and incorporation into physical theory.[170] The principle of relativity is not proved by these plausibility arguments, but it does achieve the status of a hypothesis that deserves elaboration and testing. The plausibility arguments that support the principle of relativity are so compelling that some theorists have elevated the principle to the status of a priori truth.[171] I think my brief sketch shows that its plausibility was won by long, hard experience, and that its contradictory, although extremely implausible, is not *logically impossible.*

I hope these few examples of material criteria provide some idea of the kind of plausibility argument falling under that head. The material criteria are the most important and the most interesting. Formal criteria, based upon deductive relations among hypotheses, are less frequently applicable. When they are applicable they tend to provide only maximum or minimum values. Pragmatic criteria tend to be less reliable, but not so unreliable as to be illegitimate. Material criteria supplement both other kinds and fill out the plausibility argument.

The foregoing three types of plausibility considerations—the formal criteria, the pragmatic criteria, and the material criteria—have been offered in an attempt to show how the frequency interpretation of probability can be used to approach the prior probabilities of scientific hypotheses. We have now seen how each of the major probability theories views the prior probabilities. We have seen, accordingly, that there are three distinct answers to the question of the grounds on which we can legitimately decide what kinds of hypotheses are likely to succeed. Some people would say that there are a priori principles to determine prior probabilities; they belong in the camp of the logical theorists. Others would say it is a matter of subjective predilection; they belong in the camp of the personalistic theorists. I think that both of these answers are fundamentally mistaken. There is, in my opinion, only one acceptable answer: *experience*. Those who agree in regarding experience as the only foundation for prior probabilities belong in the camp of the frequentists. This is why I remain an unregenerate frequentist against what seem to many theorists to be overwhelming difficulties. Any other answer regarding the status of prior probabilities is, to me, epistemologically unthinkable.

It may appear that my whole discussion has done very little to show how the frequentist can assign anything like precise values of prior probabilities. In fact, it may seem quite doubtful that it even makes sense to suppose exact numerical values can be established. Such an evaluation would not be too far from the truth, but fortunately it is not especially damaging to the frequency position. Numerical precision is not required, for Bayes' theorem will be applicable if we can merely judge whether or not our hypothesis is totally implausible, preposterous, and absurd. The important issue is whether the prior probability can be taken as zero for all practical purposes. If so, the hypothesis can be disqualified from further consideration—at least for the time being. Inspection of the formula reveals that a value of zero for $P(A, B)$ settles the question—in that case $P(A \cap C, B)$ is likewise zero. If however, the prior probability of the hypothesis is nonzero, the question of its posterior probability remains open. Even a very small prior probability is compatible with a very high posterior probability. Suppose we have a hypothesis with a low, but nonvanishing, prior probability. Suppose, however, that $P(A \cap \bar{B}, C)$ is also very small—i.e., the hypothesis has been confirmed by an observation whose likelihood is very small on the assumption that the hypothesis is false. Under these conditions, the posterior probability of the hypothesis can be quite large. As the personalistic theorists have emphasized, in nonextreme cases the prior probability tends to be swamped by increasing observational evidence.[172] This is a mathematical characteristic of Bayes' theorem as a formula in the uninterpreted probability calculus, so it depends in no way upon the interpretation one chooses. Thus, it is true for the frequentist as well, so that a large inaccuracy in the assessment of the prior probability may have a negligible effect upon the resulting posterior probability if a reasonable amount of confirmatory evidence is accumulated.

The Bayesian theory of scientific inference finds a very natural place for what has been known traditionally as *induction by elimination*. When Bacon sought a method to supersede induction by enumeration, he developed a system that proceeded by eliminating false hypotheses. John Stuart Mill followed him in this move. Popper's deductivism is certainly a method of elimination—of falsification—although it is controversial whether to regard it as inductive as well. The traditional objection against induction by elimination is that it is impotent in the face of an unlimited supply of possible hypotheses, for we never arrive at a unique hypothesis as a result. While this objection is valid against any form of induction by elimination that proceeds by trying to eliminate from the class of all possible hypotheses, it is not pertinent to the eliminative inference based upon Bayes' theorem. There are, as I have emphasized repeatedly, infinitely many possible hypotheses to handle any finite body of data, but it does not follow that there is any superabundance of *plausible* ones. Indeed, in practice it is often extremely difficult to think up even a couple of sensible hypotheses to explain a given problematic datum. If we put plausibility arguments—perhaps I should say "implausibility arguments"—to the purely negative task of disqualifying hypotheses with negligible prior probabilities, falsification or elimination becomes a practical approach. This is, it seems to me, the valid core of the time-honored method of induction by elimination. Like the hypothetico-deductive method, induction by elimination becomes intelligible and defensible when it is explicated in the light of Bayes' theorem, and when the indispensable role of prior probabilities is recognized.

Two basic objections have often been raised against the notion that the confirmation of scientific hypotheses could be explicated by means of the frequency interpretation. One is the objection just discussed concerning the difficulty of interpreting prior probabilities as frequencies. The second objection seems to have little to do with prior probabilities, but it becomes tractable in the Bayesian framework. It is universally recognized that the degree to which a hypothesis is confirmed depends not only upon the number of confirming instances, but also upon their variety. For instance, observations of the position of Mars confirm Newton's theory, but after a certain number of these observations each new one contributes very little to the confirmation of the theory. We want some observations of falling bodies, some observations of the tides, and a good torsion balance experiment. Any confirming instance of one of the subsequent sorts would contribute far more to the confirmation of the theory than would another observation of Mars. All of this is intuitively obvious, but—so the objection goes—the frequency interpretation cannot incorporate this basic fact about confirmation into its theory. The most it can do is *count* confirming instances; it cannot distinguish among them qualitatively.[173]

A consideration of prior probabilities seems to me to show how the problem of variety of instances can be overcome. We must note, first of all, that there is a fundamental difficulty in the very concept of variety. Any observation is different from any other in an unlimited number of ways, and any observation is similar to any other in an unlimited number of ways. It is therefore necessary to characterize similarities and differences that are relevant to confirmation. I suggest the following approach. A general hypothesis has a certain domain of applicability, and the basic idea behind the variety of instances is to test the hypothesis in different parts of its domain. It is always possible to make arbitrary partitions of the domain, but a splitting of the domain is significant only if it is not too implausible to suppose that the hypothesis holds in one part of the domain but not in another. Now we could strongly insist upon having observations of Mars on Tuesdays and Sundays as well as the other days of the week, in months whose names contain the letter "r," in years that leave a remainder of three when divided by seven, etc. However, we do not find it plausible to suppose that Newton's law holds for Mars in some of these subdomains but not in others. By contrast, it is not completely absurd to suppose that Newton's law would be suitable for bodies of astronomic dimensions located at astronomic distances from one another, but that it does not hold for smaller masses and shorter distances. Consequently, the observation of falling bodies is relevantly different, for one of the bodies involved is small even though the other (earth) is large. After observation has verified the law for falling bodies, the torsion balance experiment is very important, for it measures gravitational attraction between two bodies both of subastronomic size. The variety of instances helps us to eliminate other hypotheses, but such elimination has a point only if the alternative hypotheses being tested have nonnegligible prior probabilities.

Conclusion

The analysis of the inference by which scientific hypotheses are confirmed by observational evidence shows, I believe, that its structure is given by Bayes' theorem. This schema provides a place for the hypothetico-deductive method, showing that it is fallacious in its crude form, but that it can be made into a valid method when appropriately supplemented. Two kinds of probabilities are needed to supplement the hypothetico-deductive schema. We must assess the probability that our observational results would obtain even if the hypothesis under consideration were false. For strongest confirmation, this probability should be small. This seems a natural interpretation of Popper's methodological requirement that scientific hypotheses must be audacious and take risks. We must, in addition, assess the prior probabilities of the hypotheses we are considering. This is a reasonable interpretation of Hanson's demand for plausibility arguments.

I have argued not only that the inference schema is Bayesian, but that the probabilities that enter into the schema are to be interpreted as frequencies. It is through this interpretation, I believe, that we can keep our natural sciences empirical and objective. It enables us to show the relevance of probabilities to prediction, theory, and practical decision. Under this interpretation, induction by enumeration or a similar rule constitutes the basic inductive method for ascertainment of probabilities, but the Bayesian schema exhibits unmistakably the presence of enumerative and eliminative aspects of scientific inference, and it shows the relations between them.

If the frequency interpretation is adopted, Bayes' theorem cannot be applied until we have some experience with the success or failure of general hypotheses. On pain of infinite regress, we cannot claim that all such experience involves previous application of the Bayesian method. Instead, we must claim, logically prior to the use of Bayes' theorem some generalizations must have been established through induction by enumeration. These are hypotheses based upon crude inductive generalization, but they constitute the logical starting point. Each of them is rather shaky, owing to the childish quality of the induction by enumeration which supports it, but the more sophisticated inferences that follow can be very well founded. As evidence accumulates and further inductions are made, the results become more and more securely established.

The extensive examination of the foundations of scientific inference reveals, however, that neither induction by enumeration nor any comparable method has yet been satisfactorily justified. We cannot claim to have a well-established method for ascertaining fundamental probabilities. Hume's problem of the justification of induction remains at the foundations of scientific inference to plague those who are interested in such foundational studies.

Notes

This book is based upon five lectures in the Philosophy of Science Series at the University of Pittsburgh. The first two lectures, *Foundations of Scientific Inference:* I. *The Problem of Induction,* II. *Probability and Induction,* were presented in March 1963. The next two lectures, *Inductive Inference in Science:* I. *Hypothetico-Deductive Arguments,* II. *Plausibility Arguments,* were delivered in October 1964. The final lecture, *A Priori Knowledge,* was given in October 1965. The author wishes to express his gratitude to the National Science Foundation and the Minnesota Center for Philosophy of Science for support of research on inductive logic and probability.

1. Aristotle, *Posterior Analytics,* 100[b].
2. See, for example, Descartes, *Discourse on Method,* Pt. II, *Meditations,* or *Principles of Philosophy,* Pt. I.
3. Quoted from *Descartes Selections,* ed. R. M. Eaton (New York: Charles Scribner's Sons, 1927).
4. Francis Bacon, *Novum Organum,* aphorism xix.
5. For an excellent account of the development of calculus and the foundational problems it encountered see Carl B. Boyer, *The History of the Calculus and its Conceptual Development* (New York: Dover Publications, 1959); previously published under the title *The Concepts of the Calculus.*
6. For a simple exposition of this paradox see Bertrand Russell, *Introduction to Mathematical Philosophy* (London: Allen & Unwin, 1919), Chap. 13.
7. See Roderick M. Chisholm, "Sextus Empiricus and Modern Empiricism," *Philosophy of Science,* 8 (July 1941), 371–84.
8. I. Todhunter, *A History of the Mathematical Theory of Probability* (London, 1865).
9. John Venn, *The Logic of Chance,* 4th ed. (New York: Chelsea, 1962); 1st ed. 1866 (London).
10. *Collected Papers of Charles Sanders Peirce,* eds. C. Hartshorne and P. Weiss (Cambridge: Harvard U. Press, 1931), 6 vols.
11. John Maynard Keynes, *A Treatise on Probability* (London: Macmillan, 1952), 1st ed. 1921.

12. David Hume, *Enquiry Concerning Human Understanding,* see IV, 1.
13. Pp. 111–14.
14. For a more detailed account of the relation between deductive validity and factual content, see pp. 164–67.
15. The problem of the synthetic a priori is discussed in sec. II, 4, pp. 27–40.
16. Hume, *Human Understanding.*
17. Ibid.
18. Max Black, *Problems of Analysis* (Ithaca: Cornell U. Press, 1954), Chap. 11.
19. Ibid., pp. 196–97.
20. Lewis Carroll, "What the Tortoise Said to Achilles," in *The Complete Works of Lewis Carroll* (New York: Random House, n.d.).
21. I presented the following self-supporting argument for the counterinductive method in "Should We Attempt to Justify Induction?" *Philosophical Studies,* 8 (April 1957), 45–47. Max Black in "Self-supporting Inductive Arguments," *Models and Metaphors* (Ithaca: Cornell U. Press, 1962), Chap. 12, replies to my criticism, but he does not succeed in shaking the basic point: The counter-inductive rule is related to its self-supporting argument in precisely the same way as the standard inductive rule is related to its self-supporting argument. This is the "cash value" of claiming that the self-supporting argument is circular. Peter Achinstein, "The Circularity of a Self-supporting Inductive Argument," *Analysis,* 22 (June 1962), considers neither my formulation nor Black's answer sufficient, so he makes a further attempt to show circularity. Black's reply is found in "Self-Support and Circularity: A Reply to Mr. Achinstein," *Analysis,* 23 (December 1962). Achinstein's rejoinder is "Circularity and Induction," *Analysis,* 23 (June 1963).
22. Max Black, "The Justification of Induction," *Language and Philosophy* (Ithaca: Cornell U. Press, 1949), Chap. 3. The view he expresses in this essay, I believe, is closely related to the "probabilistic approach" I discuss in sec. II, 7, pp. 48–52.
23. Max Black, *Problems of Analysis,* p. 191.
24. Ibid., p. 206.
25. Compare Richard Bevan Braithwaite, *Scientific Explanation* (New York: Harper & Row, 1960), Chap. 8. I think the same general view is to be found in A. J. Ayer, *The Problem of Knowledge* (Baltimore: Penguin Books, 1956), p. 75. I have discussed Ayer's view in "The Concept of Inductive Evidence," *American Philosophical Quarterly,* 2 (October 1965).
26. See Braithwaite for a systematic exposition of this conception.
27. Sec. VII, pp. 108–31, "The Confirmation of Scientific Hypotheses," is devoted to a detailed analysis of this type of inference.
28. See John Patrick Day, *Inductive Probability* (New York: Humanities Press, 1961), p. 6. The nineteenth-century notion that induction is a process of discovery and the problem of whether there can be a logic of discovery are discussed in sec. VII, pp. 109–14.
29. See e.g., Karl R. Popper, *The Logic of Scientific Discovery* (New York: Basic Books, 1959), sec. 30, and Thomas S. Kuhn, *The Structure of Scientific Revolutions* (Chicago: U. of Chicago Press, 1962). A fuller discussion of the relations among such concepts as deductive validity and content is given in sec. II, 4, especially p. 33.
30. The most comprehensive statement of Popper's position is to be found in *The Logic of Scientific Discovery.* This is the English translation, with additions, of Karl R. Popper, *Logik der Forschung* (Wien, 1934).
31. "I think that we shall have to get accustomed to the idea that we must not look

upon science as a 'body of knowledge,' but rather as a system of hypotheses; that is to say, a system of guesses or anticipations which in principle cannot be justified, but with which we work as long as they stand up to tests, and of which we are never justified in saying that we know that they are 'true' or 'more or less certain' or even 'probable.'" *The Logic of Scientific Discovery,* p. 317.

32. I believe Popper openly acknowledges the nonampliative character of deduction. See "Why are the Calculi of Logic and Arithmetic Applicable to Reality," in Karl R. Popper, *Conjectures and Refutations* (New York: Basic Books, 1962), Chap. 9.
33. See *The Logic of Scientific Discovery,* Chap. 10.
34. Ibid., p. 270.
35. I shall return to Popper's methodological views in the discussion of confirmation in sec. VII. In that context I shall exhibit what I take to be the considerable valid content of Popper's account of the logic of science. See pp. 114–21.
36. For a full account of the nature of logical systems, see Alonzo Church, *Introduction to Mathematical Logic* (Princeton: Princeton U. Press, 1956), I, especially secs. 7–10 and 30.
37. Church provides a full account of interpretations of logical systems. See especially sec. 43.
38. The possibility of drawing a useful distinction between analytic and synthetic statements has been vigorously challenged by Willard Van Orman Quine in "Two Dogmas of Empiricism," *From a Logical Point of View* (Cambridge: Harvard U. Press, 1953). I am evading the problems Quine raises by supposing that we can identify logical constants, definitions, and semantic rules.
39. For an excellent discussion of this conception see Willard Van Orman Quine, "Truth by Convention," in *Readings in Philosophical Analysis,* eds. H. Feigl and W. Sellars (New York: Appleton-Century-Crofts, 1949).
40. Arthur Pap, *Introduction to the Philosophy of Science* (New York: The Free Press of Glencoe, 1962), Chap. 6, becomes confused on this point. This confusion constitutes his basis for rejection of the doctrine of conventionalism regarding the status of logical truth.
41. Lucretius, *The Nature of the Universe,* trans. Ronald Latham (Baltimore: Penguin Books, 1951), p. 31.
42. Milton K. Munitz, *Space, Time and Creation* (New York: Collier Books, 1957), Chap. 9.
43. Lucretius, pp. 66 ff.
44. René Descartes, "Of God: That He Exists," *Meditations,* III.
45. Pap, p. 97.
46. For a detailed discussion of this type of time structure, see Adolf Grünbaum, *Philosophical Problems of Space and Time* (New York: Knopf, 1963), Chap. 7.
47. See Arthur Pap, *Semantics and Necessary Truth* (New Haven: Yale U. Press, 1958), p. 103 and passim, for discussion of this and other examples.
48. Immanuel Kant, *Critique of Pure Reason* and *Prolegomena to Any Future Metaphysic.*
49. For a history of these geometrical developments, see Roberto Bonola, *Non-Euclidean Geometry* (New York: Dover Publications, 1955).
50. See Hans Reichenbach, *Philosophy of Space and Time* (New York: Dover Publications, 1958), Chap. 1, and Adolf Grünbaum, op. cit., Pt. I.
51. For an extremely readable account of the Russell-Whitehead approach, see Bertrand Russell, *Introduction to Mathematical Philosophy* (London: Allen & Unwin, 1919). For an exposition and discussion of current views on the status of

arithmetic, see Stephan Körner, *The Philosophy of Mathematics* (London: Hutchinson U. Library, 1960).

52. Norman Kemp Smith, *Immanuel Kant's Critique of Pure Reason* (London: Macmillan, 1933), p. 218.
53. David Hume, *Human Understanding,* sec. IV.
54. Ibid.
55. Ibid.
56. Ibid.
57. Ibid.
58. Ibid.
59. Ibid.
60. Wesley C. Salmon, "The Uniformity of Nature," *Philosophy and Phenomenological Research,* 14 (September 1953).
61. Bertrand Russell, *Human Knowledge: Its Scope and Limits* (New York: Simon & Schuster, 1948).
62. John Maynard Keynes, *A Treatise on Probability* (1952), Chap. 22.
63. Russell, *Human Knowledge,* p. 496.
64. Ibid.
65. Ibid.
66. Ibid., Pt. VI, Chap. 9.
67. Ibid., p. 494.
68. Arthur Burks, "The Presupposition Theory of Induction," *Philosophy of Science* 20 (July 1953) and "On The Presuppositions of Induction," *Review of Metaphysics,* 8 (June 1955).
69. Sec. III, pp. 54–56.
70. Max Black, "The Justification of Induction," in *Language and Philosophy.*
71. Among the authors who subscribe to approaches similar to this are A. J. Ayer, *Language, Truth and Logic* (New York: Dover Publications, 1952); Paul Edwards, "Russell's Doubts about Induction," *Mind,* 58 (1949), 141–63; Asher Moore, "The Principle of Induction," *J. of Philosophy,* 49 (1952), 741–58; Arthur Pap, *Elements of Analytic Philosophy* (New York: Macmillan, 1949), and *An Introduction to the Philosophy of Science;* and P. F. Strawson, *Introduction to Logical Theory* (London: Methuen, 1952).
72. I have criticized this type of argument at some length in "Should We Attempt to Justify Induction?" *Philosophical Studies,* 8 (April 1957), and in "The Concept of Inductive Evidence," *American Philosophical Quarterly,* 2 (October 1965). This latter article is part of a "Symposium on Inductive Evidence" in which Stephen Barker and Henry E. Kyburg, Jr. defend against the attack. See their comments and my rejoinder.
73. This point has enormous import for any attempt to construct an inductive justification of induction. To decide whether the fact that induction has been successful in the past is positive evidence, negative evidence, or no evidence at all begs the very question at issue.
74. As I attempted to show in "Should We Attempt to Justify Induction?" this equivocation seems to arise out of a failure to distinguish *validation* and *vindication.* This crucial distinction is explicated by Herbert Feigl, "De Principiis non Disputandum . . . ?" in *Philosophical Analysis,* ed. Max Black (Ithaca: Cornell U. Press, 1950).
75. Hans Reichenbach, *Experience and Prediction* (Chicago: U. of Chicago Press, 1938), Chap. 5, and *The Theory of Probability* (Berkeley: U. of California Press, 1949), Chap. 11.
76. Sec. V, 5, pp. 83–96.

77. Leonard J. Savage, *The Foundations of Statistics* (New York: Wiley, 1954), p. 1.
78. The most famous axiomatization is that of A. N. Kolmogorov, *Foundations of the Theory of Probability* (New York: Chelsea, 1956). This work was first published in 1933.
79. An excellent introductory account of axiomatic systems can be found in Raymond L. Wilder, *Introduction to the Foundations of Mathematics* (New York: Wiley, 1956), especially Chaps. 1 and 2. Logical systems of the kind discussed in sec. II, 4, pp. 28–31, can be presented axiomatically.
80. The axioms to be presented here are adapted from those given by Reichenbach in *The Theory of Probability,* secs. 12–14.
81. There are many excellent texts in which the basic logic of sets is presented—e.g., Wilder, Chap. 3, and Patrick Suppes, *Introduction to Logic* (Princeton: Van Nostrand, 1957), Chap. 9.
82. The Chevalier de Méré was aware that the probability of getting at least one six in three tosses of one die is less than one half while the probability of getting at least one six in four throws is over one half. He reasoned that, since the probability of a six on one throw of a die is one sixth while the probability of double six on one throw with a pair of dice is one thirty-sixth, the probability of double six in twenty-four tosses of a pair of dice should exceed one half. This is not correct. The probability is given by $1 - (35/36)^{24} = 0.4914$. In twenty-five tosses the probability exceeds one half.
83. For a simple and lucid account, see John G. Kemeny, "Carnap's Theory of Probability and Induction," in *The Philosophy of Rudolf Carnap,* ed. P. A. Schilpp (LaSalle, Illinois: Open Court, 1963), pp. 719 ff.
84. Carnap has appropriately distinguished classificatory, comparative, and quantitative concepts of probability. See *Logical Foundations of Probability* (Chicago: U. of Chicago Press, 1950; 2d ed., 1962), secs. 4, 5, and 8. The requirement of ascertainability should be taken as requiring the possibility in principle of classifying, comparative ordering, and quantitative evaluation respectively for the foregoing types of concepts. I have proceeded as if a quantitative concept is possible and desirable. For those who have serious doubts regarding quantitative probability concepts, see *Logical Foundations of Probability,* secs. 46–48.
85. Bishop Joseph Butler, *The Analogy of Religion* (1736), Preface (quoted by Carnap who quoted from Keynes; *Logical Foundations of Probability,* p. 247.)
86. Laplace is the most famous proponent of this view; see Pierre Simon, Marquis de Laplace, *A Philosophical Essay on Probabilities* (New York: Dover, 1951).
87. Excellent critical discussions are given by Richard von Mises, *Probability, Statistics and Truth,* 2d rev. English ed. (New York: Macmillan, 1957), pp. 66–81, and Hans Reichenbach, *The Theory of Probability,* secs. 68–69.
88. This contradiction is known as the "Bertrand Paradox." See von Mises, p. 77.
89. One can find formulations that sound subjective in many authors—e.g., Bernoulli, De Morgan, Keynes, Laplace, and Ramsey. On the question of whether these formulations are to be taken literally, see Carnap's illuminating discussion in *Logical Foundations of Probability,* sec. 12.
90. Ibid.
91. The classic elaboration of the logical interpretation is given in John Maynard Keynes, *A Treatise on Probability.* The most thorough and precise elaboration is to be found in Rudolf Carnap, *Logical Foundations of Probability.* For a brief and intuitively appealing presentation, see John G. Kemeny, "Carnap's Theory of Probability and Induction."
92. *Logical Foundations of Probability,* Appendix.

93. The details are given on pp. 98–103.

94. See especially Rudolf Carnap, "Replies and Systematic Expositions," sec. V, "Probability and Induction" in *The Philosophy of Rudolf Carnap*, and the "Preface to the Second Edition" of *Logical Foundations of Probability*.

95. Rudolf Carnap, *The Continuum of Inductive Methods* (Chicago: U. of Chicago Press, 1952).

96. Carnap's early answer to this question was given in "Probability as a Guide of Life," *J. of Philosophy*, 44 (1947), 141–48, and was incorporated into *Logical Foundations of Probability*, sec. 49. The adequacy of the answer was challenged by Ernest Nagel in "Carnap's Theory of Induction," in *The Philosophy of Rudolf Carnap*, and Carnap responded in "Replies and Systematic Expositions," sec. 30 in the same book.

97. See *Logical Foundations of Probability*, secs. 44, 45, and 49–51.

98. Ibid., secs. 50–51.

99. See sec. II, 7, pp. 48–52.

100. Carnap's views on the problem of justification have undergone considerable development. In "On Inductive Logic," *Philosophy of Science*, 12 (1945), 72–97, sec. 16, he endorses Reichenbach's program of attempting a pragmatic justification of induction and regards Reichenbach's argument as a positive step in the right direction, although it does not constitute a complete justification. In *Logical Foundations of Probability* (1950), sec. 41F, Carnap argues that we can establish as an analytic degree of confirmation statement that on available evidence it is probable that the degree of uniformity of nature is high. Thus, "it is reasonable for *X* to take the general decision of determining all his specific decisions with the help of inductive methods, because the uniformity of the world is probable and therefore his success in the long run is probable on the basis of his evidence . . ." (p. 181). In his "Replies and Systematic Expositions," in *The Philosophy of Rudolf Carnap*, he says, "I understand the problem of the justification of induction as the question as to what kinds of reasons can be given for accepting the axioms of inductive logic. . . . It seems to me that the reasons to be given for accepting any axiom for inductive logic have the following characteristic features . . . :

    (a) The reasons are based upon our intuitive judgements concerning inductive validity, i.e., concerning inductive rationality of practical decisions (e.g., about bets).

    Therefore:

    (b) It is impossible to give a purely deductive justification of induction.

    (c) The reasons are a priori" (p. 978).

101. Because of an isomorphism between class logic and sentential logic, Carnap's logical interpretation can be seen to satisfy the axioms presented in sec. IV, p. 59, if the arguments of the probability function are taken as sentences instead of classes. Notice that the order of the arguments must be reversed.

102. Frank Plumpton Ramsey, "Truth and Probability," *The Foundations of Mathematics and Other Logical Essays*, ed. R. B. Braithwaite (New York: Humanities Press, 1950), provides the point of departure for this theory. Its leading current exponent is Bruno de Finetti. See *Studies in Subjective Probability*, eds. H. E. Kyburg, Jr., and H. E. Smokler (New York: Wiley, 1964), for a selection of the most important papers on the personalistic interpretation.

103. See Leonard J. Savage, *The Foundations of Statistics*, Chap. 3, sec. 1, for a discussion of measurement of personal probability.

104. It is for this reason that the probability calculus cannot, by itself, provide a justification of induction.
105. Sec. VII, 3, pp. 115–31.
106. See Rudolf Carnap, "The Aim of Inductive Logic" in *Logic, Methodology and Philosophy of Science,* eds. E. Nagel, P. Suppes, and A. Tarski (Stanford U. Press, 1962).
107. The classic presentation of the frequency interpretation is John Venn, *The Logic of Chance* (see Note 9). Two of the most important twentieth-century expositions are Richard von Mises, *Probability, Statistics and Truth,* and Hans Reichenbach, *The Theory of Probability.* Not all versions of the frequency interpretation require that probability sequences be infinite. For finite frequency theories, see R. B. Braithwaite, *Scientific Explanation,* Chap. 5, and Bertrand Russell, *Human Knowledge,* Pt. V, Chap. 3.
108. This type of objection was raised against my paper, "On Vindicating Induction," in *Induction: Some Current Issues,* eds. H. E. Kyburg, Jr., and E. Nagel (Middletown, Conn.: Wesleyan U. Press, 1963) by Max Black and others. See Black's comments, the discussion, and my reply in that volume.
109. For the proof, see Reichenbach, *The Theory of Probability,* sec. 18.
110. Ibid., sec. 69.
111. The foregoing formulation differs from Reichenbach's in that his formulation mentions a degree of approximation δ, but since no method of specifying δ is given, it seems to me the rule becomes vacuous. Ibid., p. 446. It seems better to state the rule so that it equates the observed frequency and the inferred value, making it a pragmatic matter to determine, in any given case, what degree of approximation of inferred value to the true value is acceptable.

    It is also advisable, I think, to regard inductive rules as rules of inference and to avoid referring to them as rules of estimation. The main reason for this decision is that Carnap has preempted the term "estimate" in such a way that rules of estimation are not acceptance rules. Statements about estimates are analytic in just the same way as confirmation statements. Values of estimates can never be detached and asserted. It is my view that inferences concerning limits of relative frequencies yield synthetic statements that can be detached and asserted. For this reason, I always speak of *inferred values* but never of *estimates.*
112. Ibid., sec. 91.
113. Ibid., p. 447.
114. Ibid.
115. For a discussion of Reichenbach's concepts of descriptive and inductive simplicity, see *Experience and Prediction,* sec. 42.
116. See Wesley C. Salmon, "The Predictive Inference," *Philosophy of Science,* 24 (April 1957), 180–82.
117. This is the condition grossly misnamed "consistency" by Sir Ronald A. Fisher, *Statistical Methods for Research Workers* (London: Oliver & Boyd, 1954), pp. 11–12.
118. Reichenbach, *The Theory of Probability,* secs. 71–72.
119. von Mises, *Probability, Statistics and Truth,* pp. 23 ff.
120. Carnap, *Logical Foundations of Probability,* secs. 44–45.
121. Ibid., secs. 49–51.
122. Reichenbach, *Experience and Prediction,* sec. 34. "We may say: *A weight is what a degree of probability becomes if it is applied to a single case*" (p. 314, italics in original).
123. Strawson, *Introduction to Logical Theory,* p. 235.

124. Ibid.

125. A. J. Ayer, "The Conception of Probability as a Logical Relation," in *Observation and Interpretation*, ed. S. Körner (London: Butterworth's, 1957).

126. Reichenbach, *The Theory of Probability*, sec. 72, especially p. 374, and sec. 56.

127. I have no desire to minimize the problem of the short run, which is quite distinct from the problem of the single case. See my papers "The Short Run," *Philosophy of Science*, 22 (July 1955); "The Predictive Inference," *Philosophy of Science*, 24 (April 1957); and "Vindication of Induction," in *Current Issues in the Philosophy of Science*, eds. H. Feigl and G. Maxwell (New York: Holt, Rinehart & Winston, 1961).

128. Carnap, *Logical Foundations of Probability*, secs. 9, 10, 41, and 42. Hugues Leblanc, *Statistical and Inductive Probabilities* (Englewood Cliffs, N.J.: Prentice-Hall, 1962), is devoted to demonstrating the close relations between the logical concept and statistical frequencies. Henry E. Kyburg, Jr., *Probability and the Logic of Rational Belief* (Middletown, Conn.: Wesleyan U. Press, 1961), argues for a logical interpretation in which probability statements are metalinguistic statements that mention finite frequency statements.

129. Ramsey, "Truth and Probability."

130. Wesley C. Salmon, "Regular Rules of Induction," *The Philosophical Review*, 65 (July 1956); "Vindication of Induction"; and "On Vindicating Induction."

131. Wesley C. Salmon, "Vindication of Induction"; "On Vindicating Induction"; "Inductive Inference," in *Philosophy of Science: The Delaware Seminar*, II, ed. B. H. Baumrin (New York: Wiley, 1963).

132. The criterion of linguistic invariance is discussed in the articles mentioned in the preceding note. In additon, it is further discussed and elaborated in two others of my articles: "Consistency, Transitivity, and Inductive Support," *Ratio*, 71 (December 1965), and "Use, Mention, and Linguistic Invariance," *Philosophical Studies*, 17 (January 1966). It has been argued by Stephen Barker in "Comments on Salmon's 'Vindication of Induction,'" in *Current Issues in the Philosophy of Science*, and by Richard Rudner in his comments on my article and on Barker's discussion thereof, ibid., that no inductive rule can satisfy the criterion of linguistic invariance, and this includes induction by enumeration. Nelson Goodman's "grue-bleen" paradox is cited as the reason for this failure. I agree that this paradox poses a serious problem, and I have attempted to offer a solution in "On Vindicating Induction." This resolution requires that certain restrictions be placed upon the rule of induction by enumeration.

133. Carnap, *Logical Foundations of Probability*, sec. 18B and "Preface to the Second Edition."

134. Ibid., secs. 31–34.

135. Ibid., sec. 110C.

136. Kyburg, *Probability and the Logic of Rational Belief*, p. 49, discusses this difficulty, presenting a more realistic example.

137. Ian Hacking, "Salmon's Vindication of Induction," *J. of Philosophy*, 62 (May 1965).

138. John G. Kemeny, "Carnap's Theory of Probability and Induction," p. 722 f, proposes as a general *condition of adequacy* for any explication of degree of confirmation, "$c(h,e)$ is to depend only on the proposition expressed by h and e." Kemeny's discussion makes it clear that this condition, although far stronger than the criterion of linguistic invariance, does entail it. Carnap explicitly endorses this condition of adequacy; see "Replies and Systematic Expositions," p. 980.

139. Carnap, *The Continuum of Inductive Methods*, sec. 7.

140. See Kemeny, "Carnap's Theory of Probability and Induction," pp. 719–20; and Carnap, *The Continuum of Inductive Methods,* p. 43.
141. This view was espoused by W. Stanley Jevons, *The Principles of Science* (London: 1874).
142. William Whewell, *On the Philosophy of Discovery* (London: John W. Parker, 1860).
143. A word of caution is essential here. While axiom 2 of the probability calculus (p. 59) assures us that the probability of h on e is one if e entails h, the converse does not hold. See Carnap, *Logical Foundations of Probability,* sec. 58. For the analogous situation in the frequency interpretation, see Reichenbach, *The Theory of Probability,* pp. 54–55.
144. See Salmon, "Consistency, Transitivity, and Inductive Support," for further elaboration.
145. Ibid.
146. Norwood Russell Hanson, "Is There a Logic of Discovery?" *Current Issues in the Philosophy of Science.* I have tried to draw the distinction between a logic of discovery and a logic of justification in Hanson's spirit, but the contrast is drawn too dramatically. To suppose that the logic of justification is a "logic of the Finished Research Report" (p. 21) is tantamount to assuming that creative thought proceeds entirely unfettered by any critical standards until the creative work is finished. This supposition seems entirely implausible to me. I should think it is psychologically more realistic to regard discovery as a process involving frequent interplay between unfettered imaginative creativity and critical evaluation.
147. For an extremely lucid discussion of discovery and justification in deductive and inductive logic, see Carnap, *Logical Foundations of Probability,* sec. 43.
148. Church, *Introduction to Mathematical Logic,* sec. 46.
149. Hanson, "Is There a Logic of Discovery?" p. 22.
150. Ibid., p. 23.
151. Ibid.
152. Popper, *The Logic of Scientific Discovery,* p. 270.
153. I have tried to explain this elementary point in detail in *Logic* (Englewood Cliffs, N.J.: Prentice-Hall, 1963), sec. 23.
154. Pp. 60–61.
155. "Only if *the asymmetry between verification and falsification* is taken into account—that asymmetry which results from the logical relation between theories and basic statements—is it possible to avoid the pitfalls of the problem of induction." *The Logic of Scientific Discovery,* p. 265, italics in original.
156. Theorem 3, p. 60.
157. Max Born and Emil Wolf, *Principles of Optics* (New York: Pergamon Press, 1964), p. 375.
158. L. Ron Hubbard, *Dianetics: The Modern Science of Mental Healing* (Hermitage House, 1950). For an interesting account, see Martin Gardner, *Fads and Fallacies in the Name of Science* (New York: Dover, 1957), Chap. 22.
159. Popper, *The Logic of Scientific Discovery,* p. 270.
160. Popper, *Conjectures and Refutations,* p. 112.
161. "Bayesian statistics is so named for the rather inadequate reason that it has many more occasions to apply Bayes' theorem than classical statistics has." Ward Edwards, Harold Lindman, and Leonard J. Savage, "Bayesian Statistical Inference for Psychological Research," *Psychological Review,* 70 (May 1963). Under this "definition" of the Bayesian approach, Reichenbach would have to be included. "The range of application of Bayes' rule is extremely wide,

because nearly all inquiries into causes of observed facts are performed in terms of this rule. The *method of indirect evidence,* as this form of inquiry is called, consists of inferences that on closer analysis can be shown to follow the structure of the rule of Bayes." *The Theory of Probability,* p. 94. See also pp. 363, 432.

162. See Edwards, Lindman, and Savage, pp. 197–98 and 201–02.
163. Ibid., p. 208.
164. Ibid.
165. See, e.g., Ernest Nagel, *Principles of the Theory of Probability, International Encyclopedia of Unified Science,* I, no. 6 (Chicago: U. of Chicago Press, 1955), sec. 8.
166. The main ideas in the following discussion of the frequency interpretation's approach to the probability of hypotheses were presented in my paper, "The Empirical Determination of Antecedent Probabilities," read at the 1952 meeting of the American Philosophical Association, Western Division. Some of the formal details were contained in my paper, "The Frequency Interpretation and Antecedent Probabilities," *Philosophical Studies,* 4 (April 1953).

    The use of Bayes' theorem as a schema that enables the frequency interpretation to deal with the probability of hypotheses was pointed out by Reichenbach (see Note 161 above). His discussion proved quite confusing (see Nagel, *Principles of the Theory of Probability,* sec. 8). The present discussion, it is hoped, will help to clarify the position. I am deeply indebted to Reichenbach for specific suggestions concerning the interpretation of Bayes' theorem which he conveyed to me in private conversation a short time prior to his death.
167. See Gardner, *Fads and Fallacies,* pp. 28–33; and *Harpers Magazine,* 202 (June 1951), 9–11.
168. *Fads and Fallacies.*
169. For a discussion of the important distinction between two types of simplicity, descriptive simplicity and inductive simplicity, see Reichenbach, *Experience and Prediction,* sec. 42.
170. Adolf Grünbaum, *Philosophical Problems of Space and Time,* Chap. 12.
171. See, e.g., Derek F. Lawden, *An Introduction to Tensor Calculus and Relativity* (London: Methuen, 1962), p. 7.
172. See Note 162 above.
173. Nagel, *Principles of the Theory of Probability,* pp. 410 ff.

Addendum, April 1967

The Foundations of Scientific Inference, as here reprinted, is identical with the original version published in *Mind and Cosmos,* except for the correction of typographical errors. In this brief supplement I should like to do two things. First, I want to bring the references up to date by taking note of a number of important publications that have direct bearing on the issues discussed here. Second, I shall try to clarify a couple of central points on which my formulations have led to misunderstanding or confusion.

Since the writing of this study a number of significant developments have occurred, including the publication of two important monographs on aspects of probability and induction: *Logic of Statistical Inference* (Cambridge University Press, 1965), by Ian Hacking, and *The Logic of Decision* (McGraw-Hill, 1965), by R. C. Jeffrey. Both were written independently of my work; these authors did not have access to my manuscript nor I to theirs. Nevertheless, there is extensive overlap of subject matter even though there is no direct joining of issues. As the title indicates, Hacking's book deals mainly with statistical inference; consequently it is concerned with problems of the kind I have taken up in Section VI, "Inferring Relative Frequencies." He also discusses interpretations of probability, elaborating and defending a view similar to Popper's "propensity interpretation." This interpretation was not discussed in my essay because Popper's full treatment is not yet in print. It is promised in his still unpublished "Postscript: After Twenty Years," and his statements published thus far have been exceedingly brief. Popper's interpretation has a great deal in common with the frequency interpretation; especially, it takes probabilities as objective features of the world, and it construes probability statements as synthetic statements with predictive import. Thus, many of the basic arguments I have offered for the frequency interpretation might be adapted to support the propensity interpretation. Furthermore, the propensity interpretation might avoid *some* of the difficulties of the frequency interpretation, though it seems clear that it will encounter precisely the same difficulties with Hume's problem as does the frequency interpretation.

Jeffrey's book is concerned with Bayesian decision theory, so it deals mainly with the topics I have taken up in Section V, 4, "The Personalistic Interpretation," and Section VII, 3, "Bayesian Inference." One of the significant contributions of the book is a treatment of uncertain evidence. Jeffrey's arguments make it possible to refine an undesirable feature of the many systems of inductive logic in which evidence statements are simply "given," which amounts to treating them as if they were certain.

In 1966 there appeared the long-awaited festschrift for Herbert Feigl, *Mind, Matter, and Method* (University of Minnesota Press), edited by Paul K. Feyerabend and Grover Maxwell. Among several articles on induction and probability, it includes "Probability and Content Measure" by Rudolf Carnap, which greatly clarifies the relations between probability and content that are so basic to Popper's account of the logic of science (in my study see Sections II, 3, "Deductivism," and VII, 2, "Popper's Method of Corroboration," and p. 120).

During the summer of 1965, an International Colloquium in Philosophy of Science was held in London, England. The proceedings of the sessions on induction and probability are to be published in a separate volume, *The Problem of Induction* (North-Holland Publishing Co., Amsterdam), edited by Imre Lakatos. My chief contribution to the Colloquium takes up a number of issues raised in this book, and it is criticized in a most enlightening way by Ian Hacking and J. W. N. Watkins. Hacking attacks my approach to the vindication of induction and proves some interesting results concerning necessary and sufficient conditions for a justification of induction by enumeration. Watkins defends Popper's deductivism against charges of inductive pollution, and he adds delightful support to my criticisms of the probabilistic approach discussed in Section II, 7. The volume will also contain a section on acceptance rules for inductive logic (see pp. 76f, 93f, and 104f herein). The lead paper, by Henry E. Kyburg, Jr., is discussed by Yehoshua Bar-Hillel, Rudolf Carnap, and Karl Popper, among others. In my comments on Kyburg and Bar-Hillel, I attempt to reinforce the arguments for the necessity of acceptance rules.

There have also been several important reprintings. One is the useful anthology, *Probability, Confirmation, and Simplicity* (Odyssey Press, 1966), edited by Marguerite H. Foster and Michael L. Martin, which includes Hempel's classic "Studies in the Logic of Confirmation." The same article with an added postscript is reprinted, along with others of his articles on confirmation, in *Aspects of Scientific Explanation* (Free Press, 1965). Another important reprint is the second edition of Nelson Goodman's *Fact, Fiction, and Forecast* (Bobbs-Merrill, 1965). The new edition may be responsible for the large number of recent articles on Goodman's "grue-bleen" paradox (see note 132, p. 139). In an issue of *The Journal of Philosophy, LXIII,* May 26, 1966, devoted entirely to Goodman's problems, John R. Wallace in "Goodman, Logic, Induction" raises questions concerning Goodman's treatment of induction by pointing to violations of the very plausible *equivalence condition* set out by Hempel in "Studies in the Logic of Confirmation." In Wallace's article, as well as in Howard Smokler, "The Equivalence Condition," *American Philosophical Quarterly* (forthcoming, 1968), the close relation between my *criterion of linguistic invariance* and Hempel's *equivalence condition* is presented, showing the relevance of these discussions to my criterion. An answer to Wallace on behalf of Goodman by Marsha Hanen ("Goodman, Wallace, and the Equivalence Condition") appears in *The Journal of Philosophy,* LXIV, May 11, 1967.

As I have remarked in my discussion of the *criterion of linguistic invariance,* I think of it as a requirement of consistency (see pp. 101–04). Hempel also sets forth a *consistency condition* in "Studies in the Logic of Confirmation," which Carnap has shown to be untenable (*Logical Foundations of Probability,* §87). Because of a fundamental difference between our two consistency requirements, my condition is not subject to the same criticism. The difficulty for Hempel's condition arises from its connection with a qualitative notion of confirmation; since my condition is associated with a quantitative concept of inductive support or confirmation it escapes Carnap's criticism.

My preliminary attempt to explain how Bayes' theorem relates to confirmation (pp. 117f) has proved confusing to perceptive readers, largely because in that passage I vacillate between speaking of "hypotheses like *H*" and *H* itself. Later (pp. 123–25) I try to clarify the point. The original ambiguity, which was deliberate (see p. 121), arose from the fact that either manner of speaking might be correct, depending upon what interpretation of probability is adopted. For the logical and personalistic interpretations, it is straightforwardly correct to speak of the probability of *H*; for the frequency interpretation, it is not so simple. One has to distinguish, as I tried to do, between the *probability* and the *weight.* The probability pertains to a class, so we must speak of the probability of hypotheses like *H.* The weight applies to a single hypothesis, so we may speak of the weight of *H*. Since the weight is derived from the probability in an appropriate reference class, we must consider a class of hypotheses. As a result of such consideration, however, we can arrive at a weight that does characterize the individual hypothesis. All of this is a consequence of the fact that the probability (or better, weight) of a hypothesis is a special instance of the problem of the single case.

Index